Get Ready for Algebra: Grade 7
A complete workbook with lessons and problems

By Maria Miller

Copyright 2016 Maria Miller.
ISBN 978-1533161062

EDITION 12/2016

All rights reserved. No part of this workbook may be reproduced or transmitted in any form or by any means, electronic or mechanical, or by any information storage and retrieval system, without permission in writing from the author.

Copying permission: Permission IS granted for the teacher to reproduce this material to be used with students, not for commercial resale, by virtue of the purchase of this workbook. In other words, the teacher MAY make copies of the pages to be used with students.

Contents

Preface	5
Introduction	7
Helpful Resources on the Internet	8
The Order of Operations	17
Expressions and Equations	21
Properties of the Four Operations	24
Simplifying Expressions	28
Growing Patterns 1	32
The Distributive Property	35
Review 1	40
Solving Equations	42
Addition and Subtraction Equations	49
Multiplication and Division Equations	53
Word Problems	57
Constant Speed	60
Two-step Equations	67
Two-step Equations: Practice	72
Growing Patterns 2	76
Review 2	80
Answers	83
Appendix: Common Core Alignment	109

Preface

Hello! I am Maria Miller, the author of this math book. I love math, and I also love teaching. I hope that I can help you to love math also!

I was born in Finland, where I also grew up and received all of my education, including a Master's degree in mathematics. After I left Finland, I started tutoring some home-schooled children in mathematics. That was what sparked me to start writing math books in 2002, and I have kept on going ever since.

In my spare time, I enjoy swimming, bicycling, playing the piano, reading, and helping out with Inspire4.com website. You can learn more about me and about my other books at the website MathMammoth.com.

This book, along with all of my books, focuses on the conceptual side of math... also called the "why" of math. It is a part of a series of workbooks that covers all math concepts and topics for grades 1-7. Each book contains both instruction and exercises, so is actually better termed *worktext* (a textbook and workbook combined).

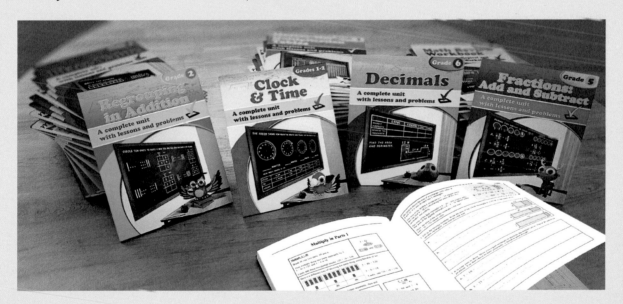

My lower level books (approximately grades 1-5) explain a lot of mental math strategies, which help build number sense — proven in studies to predict a student's further success in algebra.

All of the books employ visual models and exercises based on visual models, which, again, help you comprehend the "why" of math. The "how" of math, or procedures and algorithms, are not forgotten either. In these books, you will find plenty of varying exercises which will help you look at the ideas of math from several different angles.

I hope you will enjoy learning math with me!

Introduction

In *Get Ready for Algebra: Grade 7 Workbook*, we spend the first several lessons reviewing all of the sixth grade algebra topics and we also study some basic properties of the operations.

The main topics reviewed are the order of operations, expressions, and simplifying expressions in several different ways. The main principles are explained and practiced both with visual models and in abstract form, and the lessons contain varying practice problems that approach the concepts from various angles.

Next, students learn to solve one-step equations that involve integers. The lesson *Solving Equations* reviews the concept of an equation and how to model equations using a pan balance (scale). The basic principle for solving equations is that, when you perform the same operation on both sides of an equation, the two sides remain equal.

The workbook presents two alternatives for keeping track of the operations to be performed on an equation. The one method, writing the operation under each side of the equation, is common in the United States. The other method, writing the operation in the right margin, is common in Finland. Either is adequate, and the choice is just a matter of the personal preference of the teacher.

The introduction to solving equations is followed by a lesson on addition and subtraction equations and another on multiplication and division equations. All the equations are easily solved in only one step of calculations. The twofold goal is to make the student proficient in manipulating negative integers and also to lay a foundation for later study of more involved equations.

In the next lesson, students write equations to solve simple word problems. Even though they could solve most of these problems without using the equations, the purpose of the lesson is to make the student proficient in writing simple equations so he/she will be able to move on to more complex equations from more difficult word problems.

In the lesson *Constant Speed*, we study solving problems with distance (d), rate or velocity (v), and time (t). Students use the equivalent formulas $d = vt$ and $v = d/t$ to solve problems involving constant or average speed. They learn an easy way to remember the formula $v = d/t$ from the unit for speed that they already know, "miles per hour."

Towards the end of the workbook, we study two-step equations and practice solving them. There is also another lesson on patterns of growth, which may seem to be simply a fascinating topic, but in reality presents the fundamentals of a very important concept in algebra — that of linear functions (although they are not mentioned by that name) — and complements the study of lines in the subsequent lessons.

I wish you success in teaching math!

Maria Miller, the author

Helpful Resources on the Internet

Use these free online games and resources to supplement the "bookwork" as you see fit.

Order of operations

Otter Rush
Practice exponents in this otter-themed math game.
http://www.mathplayground.com/ASB_Otter_Rush.html

Exponents Jeopardy Game
Practice evaluating exponents, equations with exponents, and exponents with fractional bases in this interactive Jeopardy-style game.
http://www.math-play.com/Exponents-Jeopardy/exponents-jeopardy-math-game.html

Choose A Math Operation
Choose the mathematical operation(s) so that the number sentence is true. Practice the role of zero and one in basic operations or operations with negative numbers. Helps develop number sense and logical thinking.
http://www.homeschoolmath.net/operation-game.php

Order of Operations Quiz
A 10-question online quiz that includes two different operations and possibly parentheses in each question. You can also modify the quiz parameters yourself.
http://www.thatquiz.org/tq-1/?-j8f-lk-p0

The Order of Operations Millionaire
Answer multiple-choice questions that have to do with the order of operations, and win a million. Can be played alone or in two teams.
http://www.math-play.com/Order-of-Operations-Millionaire/order-of-operations-millionaire.html

Exploring Order of Operations (Object Interactive)
The program shows an expression, and you click on the correct operation (either +, −, ×, ÷ or exponent) to be done first. The program then solves that operation, and you click on the *next* operation to be performed, *etc.*, until it is solved. Lastly, the resource includes a game where you click on the falling blocks in the sequence that the order of operations would dictate.
http://www.learnalberta.ca/content/mejhm/html/object_interactives/order_of_operations/use_it.htm

Make 24 Game
Arrange the number cards, the operation symbols, and the parentheses, so that the expression will make 24.
http://www.mathplayground.com/make_24.html

Order of Operations Practice
A simple online quiz of 10 questions. Uses parentheses and the four operations.
http://www.onlinemathlearning.com/order-of-operations-practice.html

Writing expressions

Algebraic Symbolism Matching Game
Match each verbal statement with its algebraic expression.
http://www.quia.com/mc/319817.html

Who Wants to be a Hundredaire?
Try to work your way up to $100 by responding correctly to multiple-choice questions about algebraic expressions.
http://www.math-play.com/Algebraic-Expressions-Millionaire/algebraic-expressions-millionaire-game.html

Equivalent Algebraic Expressions
Practice determining whether or not two algebraic expressions are equivalent by manipulating the expressions. These problems require you to combine like terms and apply the distributive property.
https://www.khanacademy.org/math/algebra-basics/alg-basics-algebraic-expressions/alg-basics-equivalent-expressions/e/equivalent-forms-of-expressions-1

Visual Patterns
Click on the pattern to see a larger image and the answer to step 43. Can you solve the equation?
http://www.visualpatterns.org/

Expressions: Expressions and Variables Quiz
Choose an equation to match the word problem or situation.
http://www.softschools.com/quizzes/math/expressions_and_variables/quiz815.html

Translating words to Algebraic Expressions
Match the correct math expression with the corresponding English phrase, such as "7 less than a number". You can do this activity either as a matching game or as a concentration game.
https://www.quia.com/jg/1452190.html

Rags to Riches - Verbal and Algebraic Expressions
Translate between verbal and algebraic expressions in this quest for fame and fortune.
http://www.quia.com/rr/520475.html

Algebra Noodle
Play a board game against the computer while modeling and solving simple equations and evaluating simple expressions. Choose level 2 (level 1 is too easy for 7th grade).
http://www.free-training-tutorial.com/math-games/algebra-noodle.html

Matching Algebraic Expressions with Word Phrases
Five sets of word phrases to match with expressions.
http://www.mrmaisonet.com/index.php?/Algebra-Quizzes/Matching-Algebraic-Expressions-With-Word-Phrases.html

Properties of the operations

Properties of Operations at Quizlet
Includes explanations, online flashcards, and a test for the properties of operations (commutative, associate, distributive, inverse, and identity properties). The inverse and identity properties are not covered in this workbook but can be learned at the website. The identity property refers to the special numbers that do not change addition or multiplication results (0 and 1).
http://quizlet.com/2799611/properties-of-operation-flash-cards/

Commutative/associative/distributive properties matching game
Match the terms and expressions in the two columns.
http://www.quia.com/cm/61114.html?AP_rand=1554068841

Properties of Multiplication
Simple online practice about the commutative, associative, distributive, and identity properties of multiplication.
http://www.aaamath.com/pro74b-propertiesmult.html

Properties of Multiplication
Simple online practice about the commutative, associative, distributive, and identity properties of multiplication.
http://www.aaamath.com/pro74ax2.htm

Properties of the Operations Scatter Game
Drag the corresponding items to each other to make them disappear.
http://quizlet.com/763838/scatter

Associative, Distributive and Commutative Properties
Examples of the various properties followed by a simple self-test.
http://www.mathwarehouse.com/properties/associative-distributive-and-commutative-properties.php

Simplifying expressions

Simplifying Algebraic Expressions Quiz
An online quiz of 15 questions.
http://www.quia.com/quiz/1200540.html

BBC Bitesize - Simplifying Algebraic Terms
A 10-question online quiz on simplifying expressions.
http://www.bbc.co.uk/bitesize/quiz/q14530139

The distributive property

Factor the Expressions Quiz
Factor expressions such as $3x + 15$ into $3(x + 5)$.
http://www.thatquiz.org/tq-0/?-jh00-l3-p0

Distributive Property Practice
Guided practice for applying the distributive property, such as writing $-8(-7a + 10)$ as $56a - 80$.
http://www.hstutorials.net/dialup/distributiveProp.htm

Distributive Property Game
Solve questions related to the usage of the distributive property amidst playing a game. Play either a bouncing balls game or free kick soccer game, with the same questions.
http://reviewgamezone.com/games3/bounce.php?test_id=22828&title=DISTRIBUTIVE%20PROPERTY
http://reviewgamezone.com/games3/freekick.php?test_id=22828&title=DISTRIBUTIVE%20PROPERTY

Evaluate expressions

Escape Planet
Choose the equation that matches the words.
http://www.harcourtschool.com/activity/escape_planet_6/

Evaluating Expressions Quiz
Includes ten multiple-choice questions.
http://www.mrmaisonet.com/index.php?/Algebra-Quizzes/Evaluating-Expressions.html

Writing & Evaluating Expressions Quiz
This quiz has 12 multiple-choice questions and tests both evaluating and writing expressions.
http://www.quibblo.com/quiz/aWAUlc6/Writing-Evaluating-Expressions

Terms/constant/coefficient

Coefficients, Like Terms, and Constants
How to find and name the coefficients, like terms, and constants in expressions.
http://mathcentral.uregina.ca/QQ/database/QQ.09.07/h/maddie1.html

Identifying Variable Parts and Coefficients of Terms
After the explanations, you can generate exercises by pushing the button that says "new problem." The script shows you a multiplication expression, such as $-(3e)(3z)m$, and you need to identify its coefficient and variable part, effectively by first simplifying it.
http://www.onemathematicalcat.org/algebra_book/online_problems/id_var_part_coeff.htm#exercise

Tasty Term Treats
A lesson followed by a simple game where you drag terms into Toby's bowl and non-terms into the trash can.
http://mathstar.lacoe.edu/lessonlinks/menu_math/var_terms.html

Algebra - basic definitions
Clear definitions with illustrations of basic algebra terminology, including term, coefficient, constant, and expression.
http://www.mathsisfun.com/algebra/definitions.html

Equations

The Simplest Equations - Video Lessons by Maria
A set of free videos that teach the topics in this chapter — by the author herself.
http://www.mathmammoth.com/videos/prealgebra/pre-algebra-videos.php#equations

Model Algebra Equations
Model an equation on a balance using algebra tiles (tiles with numbers or the unknown x). Then, solve the equation according to instructions by placing −1 tiles on top of +1 tiles or vice versa. Includes one-step and two-step equations.
http://www.mathplayground.com/AlgebraEquations.html

One-Step Equation Game
Choose the correct root for the given equation (multiple-choice), and then you get to attempt to shoot a basket. The game can be played alone or with another student. The equations in the first game involve addition and subtraction, and in the second game (down the page) multiplication and division.
http://www.math-play.com/One-Step-Equation-Game.html

One-Step Equations Quizzes
Practice one-step equations in these timed quizzes.
http://crctlessons.com/One-Step-Equations/one-step-equations.html

http://crctlessons.com/One-Step-Equation-Test/one-step-equation-test.html

Modeling with One-Step Equations
Practice writing basic equations to model real-world situations in this interactive activity from Khan Academy.
https://www.khanacademy.org/math/pre-algebra/pre-algebra-equations-expressions/pre-algebra-equation-word-problems/e/equations-in-one-variable-1

Exploring Equations E-Lab
Choose which operation to do to both sides of an equation in order to solve one-step multiplication and division equations.
http://www.harcourtschool.com/activity/elab2004/gr6/12.html

Algebra Meltdown
Solve simple equations using function machines to guide atoms through the reactor. But don't keep the scientists waiting too long or they blow their tops.
http://www.mangahigh.com/en/games/algebrameltdown

Two-Step Equations Game
Choose the correct root for the given equation (multiple-choice), and then you get to attempt to shoot a basket. The game can be played alone or with another student.
http://www.math-play.com/Two-Step-Equations-Game.html

Two-Step Equations
Here's another five-question quiz from Glencoe that you can check yourself.
http://www.glencoe.com/sec/math/studytools/cgi-bin/msgQuiz.php4?isbn=0-07-825200-8&chapter=3&lesson=5&&headerFile=4

Solving Two-Step Equations
Type the answer to two-step-equations such as $-4y + 9 = 29$, and the computer checks it. If you choose "Practice Mode," it is not timed.
http://www.xpmath.com/forums/arcade.php?do=play&gameid=64

Two-Step Equations Word Problems
Practice writing equations to model and solve real-world situations in this interactive exercise.
https://www.khanacademy.org/math/algebra-basics/alg-basics-linear-equations-and-inequalities/alg-basics-two-steps-equations-intro/e/linear-equation-world-problems-2

Balance When Adding and Subtracting
Click on the buttons above the scales to add or subtract until you can figure out the value of x in the equation.
http://www.mathsisfun.com/algebra/add-subtract-balance.html

Algebra Four
A connect four game with equations. For this level, choose difficulty "Level 1" and "One-Step Problems".
http://www.shodor.org/interactivate/activities/AlgebraFour/

Solve Equations Quiz
A 10-question online quiz where you need to solve equations with an unknown on both sides.
http://www.thatquiz.org/tq-0/?-j102-l4-p0

Equations Level 3 Online Exercise
Practice solving equations with an unknown on both sides in this self-check online exercise.
http://www.transum.org/software/SW/Starter_of_the_day/Students/Equations.asp?Level=3

Missing Lengths
Try to figure out the value of the letters used to represent the missing numbers.
http://www.transum.org/software/SW/Starter_of_the_day/Students/Missing_Lengths.asp

Equations Level 4 Online Exercise
Practice solving equations which include brackets in this self-check online exercise.
http://www.transum.org/software/SW/Starter_of_the_day/Students/Equations.asp?Level=4

Equations Level 5 Online Exercise
This exercise includes more complex equations requiring multiple steps to find the solution.
http://www.transum.org/software/SW/Starter_of_the_day/Students/Equations.asp?Level=5

Solving Equations Quizzes
Here are some short online quizzes that you can check yourself.
http://www.glencoe.com/sec/math/studytools/cgi-bin/msgQuiz.php4?isbn=0-07-825200-8&chapter=3&lesson=5&&headerFile=4

http://www.phschool.com/webcodes10/index.cfm?wcprefix=bja&wcsuffix=0701

Rags to Riches Equations
Choose the correct root to a linear equation.
http://www.quia.com/rr/4096.html

Solve Equations Exercises
Click "new problem" (down the page) to get a randomly generated equation to solve. This exercise includes an optional graph which the student can use as a visual aid.
http://www.onemathematicalcat.org/algebra_book/online_problems/solve_lin_int.htm#exercises

Equation Word Problems Quiz
Solve word problems which involve equations and inequalities in this multiple-choice online quiz.
http://www.phschool.com/webcodes10/index.cfm?wcprefix=bja&wcsuffix=0704

Whimsical Windows - Equation Game
Write an equation for the relationship between x and y based on a table of x and y values. Will you discover the long lost black unicorn stallion?
http://mrnussbaum.com/whimsical-windows/

Speed, Time, and Distance

Distance, Speed, and Time from BBC Bitesize
Instruction, worked out exercises, and an interactive quiz relating to constant speed, time, and distance. A triangle with letters D, S, and T helps students remember the formulas for distance, speed, and time.
http://www.bbc.co.uk/bitesize/standard/maths_i/numbers/dst/revision/1/

Speed problems from Slider Math
Click on the correct speed from three choices when a distance and time are given. Often, you need to convert units in your head in order to find the correct answer.
http://www.slidermath.com/probs/Problem2.shtml

Absorb Advanced Physics - Speed
An online tutorial that teaches the concept of average speed with the help of interactive simulations and exercises.
http://www.absorblearning.com/advancedphysics/demo/units/010101.html#Describingmotion

Understanding Distance, Speed, and Time
An interactive simulation of two runners. You set their starting points and speeds, and observe their positions as the tool runs the simulation. It graphs the position of both runners in relation to time.
http://illuminations.nctm.org/Activity.aspx?id=6378

Representing Motion
A tutorial an interactive quiz with various questions about speed, time, and distance.
http://www.bbc.co.uk/schools/gcsebitesize/science/add_aqa_pre_2011/forces/represmotionrev1.shtml

Distance-Time Graphs
An illustrated tutorial about distance-time graphs. Multiple-choice questions are included.
http://www.absorblearning.com/advancedphysics/demo/units/010103.html

Distance-Time Graph
Click the play button to see a distance-time graph for a vehicle which moves, stops, and then changes direction.
http://www.bbc.co.uk/schools/gcsebitesize/science/add_aqa_pre_2011/forces/represmotionrev5.shtml

Distance versus Time Graph Puzzles
Try to move the stickman along a number line in such a way as to illustrate the graph that is shown.
http://davidwees.com/graphgame/

General

Balance Beam Activity
A virtual balance that poses puzzles where the student must think algebraically to find the weights of various figures. Includes three levels.
http://mste.illinois.edu/users/pavel/java/balance/index.html

Algebraic Reasoning
Find the value of an object based on two scales.
http://www.mathplayground.com/algebraic_reasoning.html

Algebra Puzzle
Find the value of each of the three objects presented in the puzzle. The numbers given represent the sum of the objects in each row or column.
http://www.mathplayground.com/algebra_puzzle.html

Algebraic Expressions - Online Assessment
During this online quiz you must simplify expressions, combine like terms, use the distributive property, express word problems as algebraic expressions and recognize when expressions are equivalent. Each incorrect response will allow you to view a video explanation for that problem.
http://www.mrmaisonet.com/index.php?/Algebra-Quizzes/Online-Assessment-Algebraic-Expressions.html

Stable Scales Quiz
In each picture, the scales are balanced. Can you find the weight of the items on the scales?
http://www.transum.org/software/SW/Starter_of_the_day/Students/Stable_Scales_Quiz.asp

Arithmagons
Find the numbers that are represented by question marks in this interactive puzzle.
http://www.transum.org/Software/SW/Starter_of_the_day/starter_August20.ASP

Cars
Use clues to help you find the total cost of four cars in this fun brainteaser.
http://www.transum.org/Software/SW/Starter_of_the_day/starter_July16.ASP

Algebra Quizzes
A variety of online algebra quizzes from MrMaisonet.com.
http://www.mrmaisonet.com/index.php?/Algebra-Quizzes/

Pre-algebra Quizzes
Reinforce the concepts studied in this chapter with these interactive online quizzes.
http://www.phschool.com/webcodes10/index.cfm?fuseaction=home.gotoWebCode&wcprefix=bjk&wcsuffix=0099

The Order of Operations

Let's review! Exponents are a shorthand for writing repeated multiplications by the same number.

For example, $0.9 \cdot 0.9 \cdot 0.9 \cdot 0.9 \cdot 0.9$ is written 0.9^5.

The tiny raised number is called the **exponent**.
It tells us how many times the **base** number is multiplied by itself.

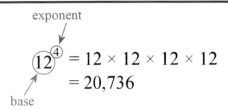

The expression 2^5 is read as "two to the fifth power," "two to the fifth," or "two raised to the fifth power."

Similarly, 0.7^8 is read as "seven tenths to the eighth power" or "zero point seven to the eighth."

The "powers of 6" are simply expressions where 6 is raised to some power: for example, 6^3, 6^4, 6^{45}, and 6^{99} are powers of 6.

Expressions with the exponent 2 are usually read as something "**squared**." For example, 11^2 is read as "eleven squared." That is because it gives us the <u>area of a square</u> with sides 11 units long.

Similarly, if the exponent is 3, the expression is usually read using the word "**cubed**." For example, 1.5^3 is read as "one point five cubed" because it is the <u>volume of a cube</u> with an edge 1.5 units long.

1. Evaluate.

   a. 4^3 **b.** 10^5 **c.** 0.1^2

   d. 0.2^3 **e.** 1^{100} **f.** 100 cubed

2. Write these expressions using exponents. Find their values.

   a. $0 \cdot 0 \cdot 0 \cdot 0 \cdot 0$ **b.** $0.9 \cdot 0.9$

   c. $5 \cdot 5 \cdot 5 \, + \, 2 \cdot 2 \cdot 2 \cdot 2 \cdot 2$

   d. $6 \cdot 10 \cdot 10 \cdot 10 \cdot 10 \cdot 10 \cdot 10 \, - \, 9 \cdot 10 \cdot 10 \cdot 10 \cdot 10$

The expression $(5 \text{ m})^3$ means that we multiply 5 meters by itself three times:

$5 \text{ m} \cdot 5 \text{ m} \cdot 5 \text{ m} = 125 \text{ m}^3$

Notice that $(5 \text{ m})^3$ is different from 5 m^3. The latter has no parentheses, so the exponent (the little 3) applies only to the unit "m" and not to the whole quantity 5 m.

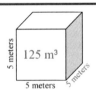

3. Find the value of the expressions.

   a. $(2 \text{ cm})^3$ **b.** $(11 \text{ ft})^2$ **c.** $(1.2 \text{ km})^2$

4. Which expression from the right matches with (a) and (b) below?

   a. The volume of a cube with sides 2 cm long.

                                             (i) 8 cm^3 (ii) $(8 \text{ cm})^3$ (iii) 2 cm^3

   b. The volume of a cube with sides 8 cm long.

The Order of Operations (PEMDAS)	You can remember PEMDAS with the silly mnemonic *Please Excuse My Dear Aunt Sally.*
1) Solve what is within parentheses (**P**).	
2) Solve exponents (**E**).	Or make up your own!
3) Solve multiplication (**M**) and division (**D**) from left to right.	
4) Solve addition (**A**) and subtraction (**S**) from left to right.	

5. Find the value of these expressions.

a. $120 - (9 - 4)^2$	c. $4 \cdot 5^2$	e. $10 \cdot 2^3 \cdot 5^2$
b. $120 - 9 - 4^2$	d. $(4 \cdot 5)^2$	f. $10 + 2^3 \cdot 5^2$
g. $(0.2 + 0.3)^2 \cdot (5 - 5)^4$	h. $0.7 \cdot (1 - 0.3)^2$	i. $20 + (2 \cdot 6 + 3)^2$

Example 1. Solve $(10 - (5 - 2))^2$.

First solve what is within the *inner* parentheses: $5 - 2 = 3$. We get $(10 - 3)^2$.

The rest is easy: $(10 - 3)^2 = 7^2 = 49$.

Example 2. Simplify $2 + \dfrac{1+5}{6^2}$.

Remember, the fraction line works like parentheses as a grouping symbol, grouping both what is above the line and also what is below it. First solve $1 + 5$, then the exponent.

$$2 + \frac{1+5}{6^2} = 2 + \frac{6}{6^2} = 2 + \frac{1}{6} = 2\frac{1}{6}$$

6. Find the value of these expressions.

a. $(12 - (9 - 4)) \cdot 5$	c. $(10 - (8 - 5))^2$
b. $12 - (9 - (4 + 2))$	d. $3 \cdot (2 - (1 - 0.4))$

7. Find the value of these expressions.

a. $\dfrac{4 \cdot 5}{2} \cdot \dfrac{9}{3}$	b. $\dfrac{4 \cdot 5}{2} + \dfrac{9}{3}$	c. $\dfrac{4 + 5}{2} + \dfrac{9}{3 - 1}$

Expressions written using the ÷ symbol can be rewritten using the fraction line.
This usually makes them easier to read.

Example 3. In the expression $2 + 5 \cdot 2 \div 4 \cdot 10$, the division is by 4. This means that when written using the fraction line, only 4 goes in the denominator.

The expression becomes $2 + \dfrac{5 \cdot 2}{4} \cdot 10$

Here is how to simplify it:

$2 + \dfrac{5 \cdot 2}{4} \cdot 10$

$= 2 + \dfrac{10}{4} \cdot 10$

$= 2 + \dfrac{100}{4}$ (or $2 + 2.5 \cdot 10$)

$= 2 + 25 = 27$

Example 4. Rewrite the expression $2 \div 4 \cdot 3 \div (7 + 2) + 1$ using the fraction line.

Now there are *two* divisions: the first by 4 and the second by $(7 + 2)$. This means we will use *two* fractions in the expression.

It is written as $\dfrac{2}{4} \cdot \dfrac{3}{7 + 2} + 1$.

To simplify it, first calculate $7 + 2$, remembering that the fraction line implies parentheses around both the numerator and the denominator.

We get $\dfrac{2}{4} \cdot \dfrac{3}{9} + 1$.

Reducing the fractions, 2/4 equals 1/2, and 3/9 equals 1/3.

We get $\dfrac{1}{2} \cdot \dfrac{1}{3} + 1 = \dfrac{1}{6} + 1 = 1\dfrac{1}{6}$.

8. Rewrite each expression using a fraction line, then simplify. Compare the expression in the top row with the one below it. *Hint: Only what comes right after the "÷" sign goes into the denominator.*

a. $2 \div 5 \cdot 4$	b. $16 \div (2 + 6) \cdot 2$	c. $4 + 1 \div 3 + 2$
d. $2 \div (3 \cdot 4)$	e. $5 \div 9 \cdot 3 + 1$	f. $(1 + 3) \div (4 + 2)$

9. Find the value of these expressions. (Give your answer as a fraction or mixed number, not as a decimal.)

a. $\dfrac{9^2}{9} \cdot 6$	b. $\dfrac{2^3}{3^2}$	c. $\dfrac{(5-3) \cdot 2}{8 - 1 + 2} + 10$

10. Evaluate the expressions. (Give your answer as a fraction or mixed number, not as a decimal.)

a. $2x^2 - x$, when $x = 4$	**b.** $3s - 2t + 8$, when $s = 10$ and $t = 5$
c. $\dfrac{x^2}{x+1}$, when $x = 3$	**d.** $\dfrac{x+1}{x-1}$, when $x = 11$
e. $\dfrac{a+b}{b} + 2$, when $a = 1$ and $b = 3$	**f.** $\dfrac{n^2 + 2n}{n+3}$, when $n = 5$

11. Write a single mathematical expression ("number sentence") for each situation. Don't write just the answer.

a. You buy n hats for $4 each and m scarves for $6 each. Write an expression for the total cost. *cost* =	**b.** You have x pennies and y dimes in your pocket. What is their total value in cents? *value in cents* =
c. Molly and Mike share 10 cookies between them. Molly gets t cookies. Write an expression for how many cookies Mike gets. *Mike's cookies* =	**d.** Heather earns $11 per hour. Write an expression for how much she earns in n hours. *earnings* =
e. The club has 81 members, and 2/3 of them are girls. Write an expression for the number of girls. *girls* =	**f.** The club has n members, and 2/3 of them are girls. Write an expression for the number of girls. *girls* =
g. The price of a $60 book is discounted by 1/10. Write an expression for the current price. *price* =	**h.** The price of a book costing x is discounted by 1/10. Write an expression for the discounted price. *price* =
i. The altitude of a triangle is 3 and its base is b. Write an expression for its area (A). A =	**j.** The edge of a cube is c units long. Write an expression for its volume (V). V =

Expressions and Equations

Expressions in mathematics consist of: • numbers; • mathematical operations $(+, -, \cdot, \div, \text{exponents})$; • and letter variables, such as $x, y, a, T,$ and so on. Note: Expressions do *not* have an "equals" sign! **Examples of expressions:** $\quad 5 \quad\quad \dfrac{xy^4}{2} \quad\quad T-5$	An **equation** has two expressions separated by an equals sign: $\quad$(expression 1) = (expression 2) **Examples:** $\quad 0 = 0 \quad\quad 2(a-6) = b$ $\quad 9 = -8 \quad\quad \dfrac{x+3}{2} = 1.5$ (a false equation)
What do we do with expressions? We can find the *value* of an expression (*evaluate* it). If the expression contains variables, we cannot find its value unless we know the value of the variables. For example, to find the value of the expression $2x$ when x is $3/7$, we simply substitute $3/7$ in place of x. We get $2x = 2 \cdot 3/7 = 6/7$. Note: When we write $2x = 2 \cdot 3/7 = 6/7$, the equals sign is *not* signaling an equation to solve. (In fact, we already know the value of x!) It is simply used to show that the value of the expression $2x$ here is the same as the value of $2 \cdot 3/7$, which is in turn the same as $6/7$.	**What do we do with equations?** If the equation has a variable (or several) in it, we can try to *solve* the equation. This means we find the values of the variable(s) that make the equation true. For example, we can solve the equation $0.5 + x = 1.1$ for the unknown x. The value 0.6 makes the equation true: $0.5 + 0.6 = 1.1$. We say $x = 0.6$ is the **solution** or the **root** of the equation.

1. This is a review. Write an expression.

   a. $2x$ minus the sum of 40 and x.

   b. The quantity 3 times x, cubed.

   c. s decreased by 6

   d. five times b to the fifth power

   e. seven times the quantity x minus y

   f. the difference of t squared and s squared

   g. x less than 2 cubed

   h. the quotient of 5 and y squared

   i. 2 less than x to the fifth power

   j. x cubed times y squared

   k. the quantity $2x$ plus 1 to the fourth power

   l. the quantity x minus y divided by the quantity x squared plus one

> To read the expression $2(x + y)$, use the word ***quantity***:
> "two times the quantity x plus y."
>
> There are other ways, as well, just not as common:
>
> "two times the sum of x and y," or
> "the product of 2 and the sum x plus y."

Some equations are *true*, and others are *false*. For example, 0 = 9 is a false equation.

Some equations are neither. The equation $x + 1 = 7$ is neither false nor true in itself. However, if x has a specific value, then we can tell if the equation becomes true or false.

Indeed, solving an equation means finding the values of the variables that make the equation *true*. The solutions of the equation can also be called its **roots**.

Example. Find the root of the equation $20 - 2y^2 = 2$ in the set {0, 1, 2, 3, 4}.

Try each number from the set, checking to see if it makes the equation true:

$20 - 2 \cdot 0^2 = 20 \neq 2$ $20 - 2 \cdot 2^2 = 12 \neq 2$ $20 - 2 \cdot 4^2 = -12 \neq 2$
$20 - 2 \cdot 1^2 = 18 \neq 2$ $20 - 2 \cdot 3^2 = 2$

So, in the given set, the only root of the equation is 3.

2. Write an equation. Then solve it.

	Equation	Solution
a. 78 decreased by some number is 8.		
b. The difference of a number and 2/3 is 1/4.		
c. A number divided by 7 equals 3/21.		

3. **a.** Find the root(s) of the equation $n^2 - 9n + 14 = 0$ in the set on the right.

   1   10   3
     6   2   7

   b. Find the root(s) of the equation $9x - 5 = 2x$ in the set on the right.

   1/5   5/7
     7/9

4. Which of the numbers 0, 1, 3/2, 2 or 5/2 make the equation $\dfrac{y}{y-1} = 3$ true?

5. **a.** Ann is 5 years older than Tess, and Tess is n years old.
   Write an <u>expression</u> for Ann's age.

   b. Let A be Alice's age and B be Betty's age.
   Find the <u>equation</u> that matches the sentence
   "Alice is 8 years younger than Beatrice."

   A = 8 − B   A = B − 8   B = A − 8

   Hint: Give the variables some test values.

6. **a.** In a bag of blocks, there are twice as many red blocks as there are blue blocks, and three times as many green blocks as blue blocks. Let's denote the number of blue blocks with *x*. Write an expression for the amount of red blocks.

   b. Write another expression for the amount of green blocks.

7. **a.** Timothy earns *s* dollars in a month. He pays 1/5 of it in taxes and gets to keep 4/5 of it. Write an expression for the amount Timothy gets to keep.

   b. Write another, different expression for the amount Timothy gets to keep.
   (Hint: If you used a fraction in a., use a decimal now, or vice versa).

8. Circle the equation that matches the situation. *Hint: Give the variable(s) some value(s) to test the situation.*

   a. The price of a phone is discounted by 1/4, and now it costs $57.

   | $\frac{p}{4} - \$57$ | $\frac{3p}{4} = \$57$ | $\frac{4p}{3} = \$57$ | $\frac{p}{3} = 4 \cdot \$57$ | $p - 1/4 = \$57$ |

   b. Matt bought three computer mice for $25 each and five styluses for *p* dollars each. He paid a total of $98.

   | $25 + 5p = 98$ | $3 \cdot 25 + 5p = 98$ | $3p + 125 = 98$ |
   | $3 \cdot 25 \cdot 5p = 98$ | $3 \cdot 25 + p = 98$ | $5p + 75 = 98$ |

   c. Jeremy sells fresh-squeezed orange juice for *x* per glass. Today he has discounted the glass of juice by $1. A customer buys three glasses, and the total comes to $5.40.

   | $3(x - \$1) = \$5.40$ | $3x - \$1 = \$5.40$ |
   | $x - \$1 = 3 \cdot \$5.40$ | $3(x - 0.1) = \$5.40$ |

Puzzle Corner

Here is a very strange equation: $n = n$
If you think about it, you can put *any* number in place of *n*, and the equation will be true!

For example, if *n* is 5, we get 5 = 5 (a true equation). This equation has an **infinite number of solutions**—any number *n* will make it true!

Find the equations below that also have an infinite number of solutions.

$1 + x = 2 + x$ $4 + c + 1 = 2 + 3 + c$ $2y - 10 = y + y$

$3z - 1 = z - 1 + z + z$ $6 + 2n + 3n = n + 6$ $b \cdot b = 0$

Properties of the Four Operations

In this lesson, we will look at some special properties of the basic operations. You already know them. In fact, you have been using them since you first learned to add! But this time, we will name the properties and study them in detail.

1. Addition is **commutative**.

If your father commutes to work, he changes where he is at from home to work and back. In math, when a and b commute, they change places. This means that you can change the order of the addends when you add two numbers. In symbols: $a + b = b + a$.

In other words: When adding 2 numbers, you can change their order.

2. Addition is **associative**.

When you associate with people, you group yourself with them. In math, when a and b associate, they are grouped together. The associative property says that when adding three numbers, it does not matter if you begin by adding the first two or the last two. In symbols: $(a + b) + c = a + (b + c)$.

(What about adding a and c first? Would that work?)

Then we have the identical properties for multiplication.

3. Multiplication is **commutative**.

When multiplying two numbers, you can change their order. In symbols: $ab = ba$.

4. Multiplication is **associative**.

When multiplying three numbers, it does not matter if you start with the first two or if you start with the last two. In symbols: $(ab)c = a(bc)$.

(Could you even start by multiplying a and c first? Would that work?)

1. Are the two expressions in each box equivalent? That is, do they have the same value for any value of c? Give c some test values to check.

a. $c + 5$	b. $c - 5$	c. $c \div 6$	d. $5c$
$5 + c$	$5 - c$	$6 \div c$	$c \cdot 5$

2. Is subtraction commutative? In other words, will $a - b$ always have the same value as $b - a$, no matter what values we give to a and b? Explain your reasoning.

3. Is division commutative? Does $a \div b$ always have the same value as $b \div a$ for any numbers that we might use for a and b? Explain your reasoning.

4. Are the expressions equal, no matter what value n has? Give n some test values to check.

a. $(n + 2) + 5$	b. $(n \cdot 2) \cdot 5$	c. $(n - 7) - 3$
$n + (2 + 5)$	$n \cdot (2 \cdot 5)$	$n - (7 - 3)$

5. **a.** Name the property of arithmetic illustrated by (a) above.

   b. Name the property of arithmetic illustrated by (b) above.

6. Are the expressions equal, no matter what values n and m have? If so, you don't need to do anything else. If not, provide a counterexample: specific values of n and m that show the expressions do NOT have the same value.

a. $n - 2 - m$	b. $(2n + 1) \cdot 5$
$n + (2 - m)$	$5 \cdot (1 + 2n)$
Not equal. For example when $n = 5$ and $m = 1$, we get $n - 2 - m = 5 - 2 - 1 = \underline{2}$, but $n + (2 - m) = 5 + (2 - 1) = \underline{6}$.	
c. $(n - 2) \cdot m$	d. $a + 2b$
$m(2 - n)$	$b + 2a$

7. Is subtraction associative? In other words, is it true that $(a - b) - c$ has the same value as $a - (b - c)$, no matter what values a, b, and c get? Explain your reasoning.

8. Is division associative? In other words, is it true that $(a \div b) \div c$ has the same value as $a \div (b \div c)$, no matter what values a, b, and c get? Explain your reasoning.

Since addition is both commutative and associative, it follows that **we can add a list of numbers in any order we choose.** Of course, you already knew that!

(Optional) Here is a proof that, when you add three numbers, you can start with the first and last numbers.

Consider the sum of three numbers $a + b + c$.

1. Because addition is <u>commutative</u>, we can switch the order of a and b.

   Thus $b + a + c$ has the same value as $a + b + c$.

2. Because addition is associative, $b + a + c = b + (a + c)$. So instead of proceeding from left to right and adding the first two numbers first, we can add the last two numbers first.

3. Therefore we can indeed calculate the sum $a + c$ first because it is inside the parentheses.

The same is true of multiplication: **you can multiply a list of numbers in any order you choose.**

We can use these properties of operations to simplify expressions.

Example 1. Simplify $5 + a + b + 7 + a$.

Here is a model for this expression: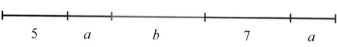

Because we can add in any order, we can add $5 + 7$ to get 12. Moreover, we can add $a + a$ and write that as $2a$. So $5 + a + b + 7 + a$ simplifies to $2a + b + 12$. That is as simplified as it can get.

Note: It's customary to write the terms with variables in alphabetical order and put the constant term (here, the "12") last.

9. Write an expression from the illustration and simplify it.

a.
$x \quad x \quad x \quad x \quad x$

b.
$a \quad a \quad b \quad b \quad a$

c.
$s \quad s \quad 15 \quad s \quad s$

d.
$v \quad 11 \quad v \quad 16 \quad v \quad t$

10. Simplify the expressions.

a. $5 + v + 8 + v + v$	b. $e + e + 9 + e + 28 + e$	c. $2v + v$
d. $5a + 8a$	e. $8 + 6a + 5b + 3b + 9a$	f. $10t + s + 2 + s + 3s$

> **Example 2.** Simplify $x \cdot x \cdot 4 \cdot y \cdot y \cdot y \cdot y$.
>
> The part $x \cdot x$ can be written as x^2, and the part $y \cdot y \cdot y \cdot y$ as y^4.
> Those are multiplied, so in total we get $x^2 \cdot 4 \cdot y^4$.
>
> Since the multiplication sign is usually omitted from between variables, and the constant 4 is usually written in front of the term, we would write this as $4x^2y^4$.

11. Simplify the expressions.

a. $a \cdot a \cdot 7$	**b.** $2 \cdot s \cdot s \cdot 8$	**c.** $a \cdot a \cdot d \cdot d \cdot d \cdot d$

Error alert!

Multitudes of algebra students have confused $b \cdot b \cdot b$ with $b + b + b$ and written $b \cdot b \cdot b = 3b$.

That is NOT true! Make sure you understand and remember the reason why:

Just as $b + b + b$ is repeated addition, for which the shortcut is to multiply: $b + b + b = 3b$; so, too, $b \cdot b \cdot b$ is repeated multiplication, for which the shortcut is to use an exponent: $b \cdot b \cdot b = b^3$.

d. $y \cdot x \cdot x \cdot y \cdot 2 \cdot y \cdot x$	**e.** $d + a + a + d$	**f.** $z + z + z + 8 + z$
g. $y \cdot y \cdot y \cdot 8 \cdot t \cdot t$	**h.** $b \cdot b \cdot 9 \cdot b \cdot 3 \cdot 1 \cdot b$	**i.** $2s + s + t + 3s + 2$

12. Are the expressions equal, no matter what values x and y have? If yes, you don't need to do anything else. If not, provide a counterexample.

a. $\dfrac{5}{x}$ $\dfrac{x}{5}$	**b.** $x + \dfrac{y}{2}$ $y + \dfrac{x}{2}$
c. $\dfrac{x+y}{2}$ $\dfrac{x}{2} + \dfrac{y}{2}$	**d.** $\dfrac{x}{x}$ $\dfrac{y}{y}$

13. Summary. Write "yes" or "no" to indicate if the operation is commutative or associative. Include examples or comments if you want to.

Operation	Commutative?	Associative?	Optional notes/examples
addition			
subtraction			
multiplication			
division			

Simplifying Expressions

Example. Simplify $2x \cdot 4 \cdot 5x$.

Notice, this expression contains only multiplications (because $2x$ and $5x$ are also multiplications).

Since we can multiply in any order, we can write this expression as $2 \cdot 4 \cdot 5 \cdot x \cdot x$.

Now we multiply 2, 4, and 5 to get 40. What is left to do? The part $x \cdot x$, which is written as x^2.

So, $2x \cdot 4 \cdot 5x = 40x^2$.

Note: The equals sign used in $2x \cdot 4 \cdot 5x = 40x^2$ signifies that the two expressions are equal no matter what value x has. That equals sign does not signify an equation that needs to be solved.

Similarly, we can simplify the expression $x + x$ and write $2x$ instead. That whole process is usually written as $x + x = 2x$.

Again, the equals sign there does not indicate an equation to solve, but just the fact that the two expressions are equal. In fact, if you think of it as an equation, *any* number x satisfies it! (Try it!)

1. Simplify the expressions.

a. $p + 8 + p + p$	**b.** $p \cdot 8 \cdot p \cdot p \cdot p$	**c.** $2p + 4p$
d. $2p \cdot 4p$	**e.** $5x \cdot 2x \cdot x$	**f.** $y \cdot 2y \cdot 3 \cdot 2y \cdot y$

2. Write an expression for the area and perimeter of each rectangle.

a. (square with side x)

area =

perimeter =

b. (square with side x, wait — rectangle with width x and height x)

area =

perimeter =

c. (rectangle $4x$ by $2x$)

area =

perimeter =

d. (square $3x$ by $3x$)

area =

perimeter =

3. **a.** Sketch a rectangle with sides $2b$ and $7b$ long.

   b. What is its area?

   c. What is its perimeter?

4. **a.** The perimeter of a rectangle is $24s$.
   Sketch one such rectangle.

   What is its area?
   Hint: There are many possible answers.

   b. Find the area and perimeter of your rectangle
   in (a) if s has the value 3 cm.

5. **a.** Which expression below is for an area of a rectangle? Which one is for a perimeter?

   $4a + 4b$ $2a \cdot 2b$

   b. Sketch the rectangle.

6. **a.** Find the value of the expressions $3p$ and $p + 3$ for different values of p.

Value of p	3p	p + 3
0		
0.5		
1		
1.5		
2		
2.5		
3		
3.5		
4		

   b. Now, look at the table. Can you tell which is larger, $3p$ or $p + 3$?

Some review! In algebra, a **term** is an expression that consists of numbers and/or variables that are multiplied together. A single number or variable is also a term.

Examples.
- $2xy$ is a term, because it only contains multiplications, a number, and variables.
- $(5/7)z^3$ is a term. Remember, the exponent is a shorthand for repeated multiplication.
- Addition and subtraction separate the individual terms in an expression from each other. For example, the expression $2x^2 - 6y^3 + 7xy + 15$ has four terms, separated by the plus and minus signs.
- $s + t$ is *not* a term, because it contains addition. Instead, it is a sum of *two* terms, s and t.

The number by which a variable or variables are multiplied is called a ***coefficient***.

Examples.
- The term $0.9ab$ has the coefficient 0.9.
- The coefficient of the term m^2 is 1, because you can write m^2 as $1 \cdot m^2$.

If the term is a single number, such as 7/8, we call it a ***constant***.

Example. The expression $1.5a + b^2 + 6/7$ has three terms: $1.5a$, b^2, and $6/7$. The last term, $6/7$, is a constant.

7. Fill in the table.

Expression	The terms in it	Coefficient(s)	Constants
$(5/6)s$			
w^3			
$0.6x + 2.4y$			
$x + 3y + 7$			
$p \cdot 101$			
$x^5y^2 + 8$			

The two terms in the expression $2x^2 + 5x^2$ are **like terms**: they have the same variable part (x^2). Because of that, we can add the two terms to simplify the expression. To do that, simply add the coefficients 2 and 5 and use the same variable part: $2x^2 + 5x^2 = 7x^2$. It is like adding 2 apples and 5 apples.

However, you cannot add (or simplify) $2x + 7y$. That would be like adding 2 apples and 7 oranges.

Example. Simplify $6x - x - 2x + 9x$. The terms are like terms, so we simply add or subtract the coefficients: $6 - 1 - 2 + 9 = 12$ and tag the variable part x to it. The expression simplifies to $12x$.

8. Simplify the expressions.

a. $5p + 8p - p$	**b.** $p^2 + 8p^2 + 3p^2$	**c.** $12a^2 - 8a^2 - 3a^2$

9. Write an expression for the total area.

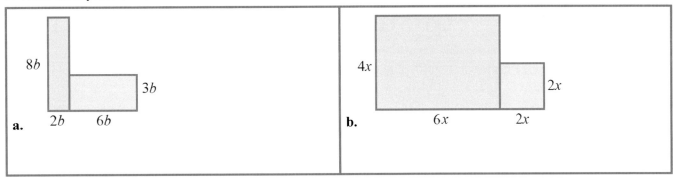

In the following problems, write an expression for part (a), and then for part (b) write an equation and solve it. Don't skip writing the equation, even if you can solve the problem without it, because we are practicing writing equations! You don't have to use algebra to solve the equations—you can solve them in your head or by guessing and checking.

10. **a.** The length of a rectangle is 4 meters and its width is w. What is its perimeter? Write an expression.

   b. Let's say the perimeter has to be 22 meters. How wide is the rectangle then? Write an *equation* for this situation, using your expression from (a).

   Remember, you do not have to use algebra to solve the equation—you can solve it in your head or by "guess and check." But do write the equation.

11. **a.** Linda borrows six books from the library each week, and her mom borrows two. How many books, in total, do both of them borrow in w weeks? Write an expression.

   b. How many weeks will it take them to have borrowed 216 books? Write an equation.

12. **a.** Alice buys y containers of mints for $6 apiece. A fixed shipping cost of $5 is added to her order. What is her total cost? Write an expression.

   b. Alice's total bill was $155. How many containers of mints did she buy? Write an equation.

Puzzle Corner

a. What is the total value, in cents, if Ashley has n dimes and m quarters? Write an expression.

b. The total value of Ashley's coins is 495 cents. How many dimes and quarters can she have?
Hint: Make a table to organize the possibilities.

Growing Patterns 1

This is a pattern of squares. In each step, the side of the square grows by 1 flower.

Step   1   2   3   4   5

Draw steps four and five.

How many flowers will there be in step 13?

How many flowers will there be in step n?
(If you get stuck, the answer is at the bottom of the page, but try it on your own first.)

Here is another pattern:

Step   1   2   3   4   5

Draw steps four and five.

How do you see this pattern growing? (There's more than one way to look at it!)

How many flowers will there be in step 39?

What about in step n?

(The answers are in the next box, but stop here to try the problem yourself first.)

---

1. One way is to see the pattern as an L-shape. This means in step 39 there are 39 flowers horizontally and 39 vertically, but with 1 flower overlapping, so the overlapping flower needs to be subtracted. In total, there are $39 + 39 - 1 = 77$ flowers in step 39.

   Similarly, in step n, there are $n + n - 1$ flowers, or $2n - 1$ flowers.

2. Another way to see it is as a square from which is subtracted a smaller square. This means that, in step 39, we have a square that is $39 \cdot 39$, and then a square $38 \cdot 38$ is taken away. $39 \cdot 39 - 38 \cdot 38 = 1521 - 1444 = 77$.

   In step n the length of the side of the square is n, so its area is n^2. The square that is "taken away" has a side $n - 1$ long (one less than the side of the bigger square). Its area is $(n-1)^2$.

   Then we subtract these two areas to find that there are $n^2 - (n-1)^2$ flowers in each step.

   Of course, this expression is more complicated than $2n + 1$, but believe it or not, the two *are* equivalent!

Answer for the pattern at the top of this page: There are n^2 flowers in step n.

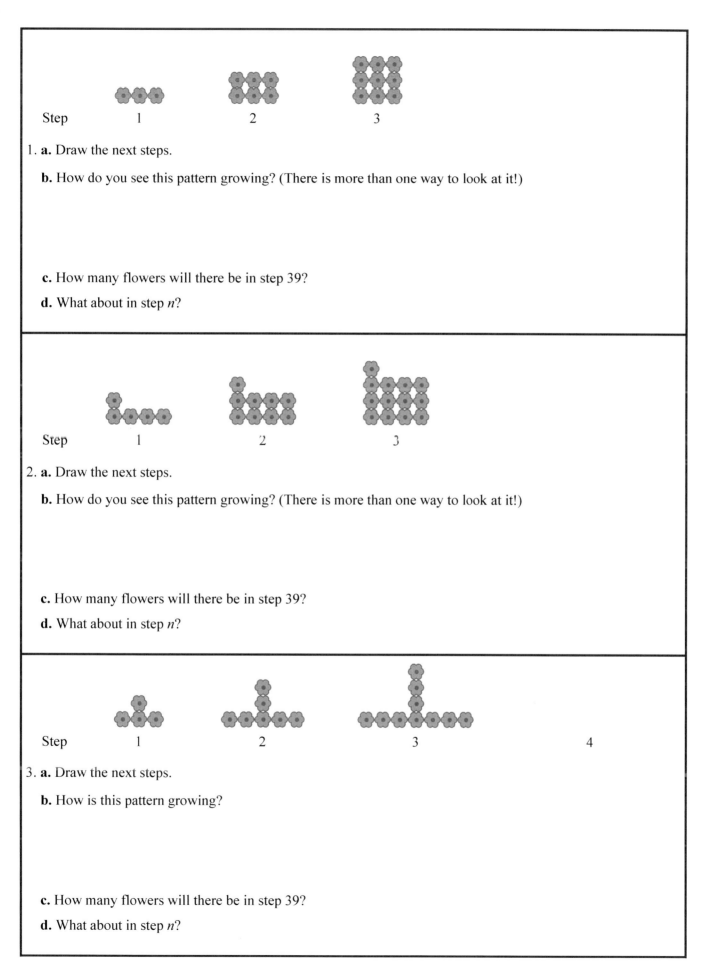

Step 1 2 3

1. a. Draw the next steps.
   b. How do you see this pattern growing? (There is more than one way to look at it!)
   c. How many flowers will there be in step 39?
   d. What about in step *n*?

Step 1 2 3

2. a. Draw the next steps.
   b. How do you see this pattern growing? (There is more than one way to look at it!)
   c. How many flowers will there be in step 39?
   d. What about in step *n*?

Step 1 2 3 4

3. a. Draw the next steps.
   b. How is this pattern growing?
   c. How many flowers will there be in step 39?
   d. What about in step *n*?

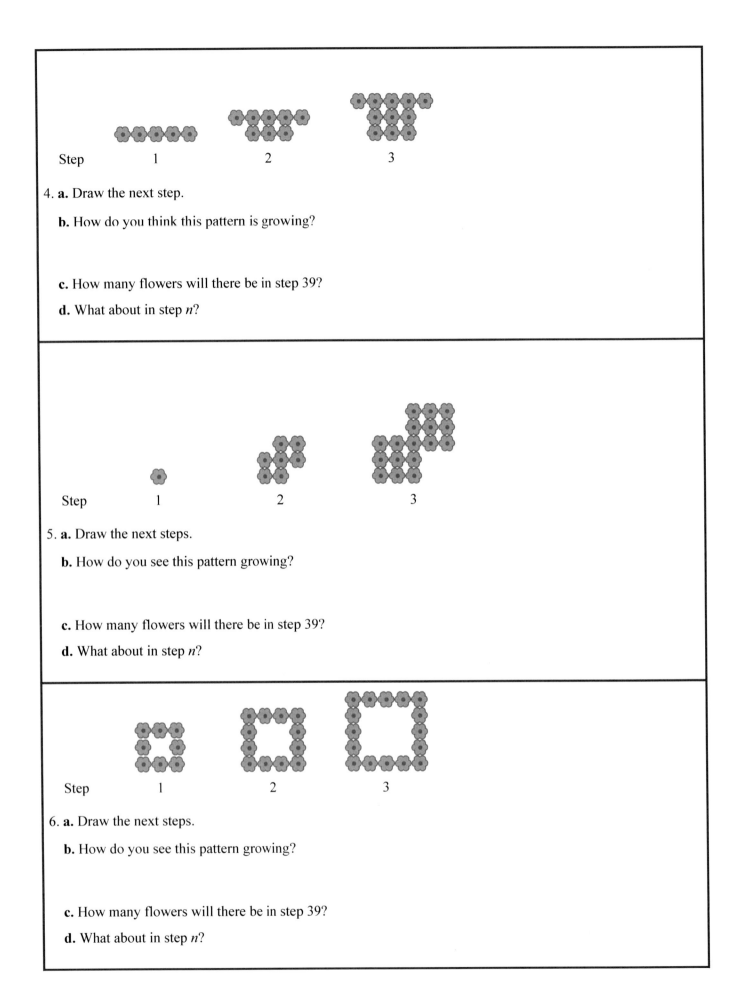

The Distributive Property

The **distributive property** states that $a(b + c) = ab + ac$ for any numbers a, b, and c.

It says we can *distribute* multiplication over addition. This means that instead of multiplying a times the sum $b + c$, we can multiply the numbers b and c separately by a and add last.

Example 1. The expression $20(x + 5)$ is equal to $20x + 20 \cdot 5$, which simplifies to $20x + 100$.

Notice what happens: Each term in the sum $(x + 5)$ gets multiplied by the factor 20! Graphically:

$$20(x + 5) = 20x + 20 \cdot 5$$

Example 2. To multiply $2a(3 + c)$ using the distributive property, we need to multiply **both** 3 and c by $2a$:

$$2a(3 + c) = 2a \cdot 3 + 2a \cdot c$$

Lastly, we simplify: $2a \cdot 3$ simplifies to $6a$, and of course we can write $2a \cdot c$ without the multiplication sign:

$$2a \cdot 3 + 2a \cdot c = 6a + 2ac$$

Here is a way to **model the distributive property using line segments.**

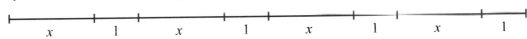

The model shows a pattern of line segments of lengths x and 1 repeated four times.
In symbols, we write $4 \cdot (x + 1)$.

However, it is easy to see that the total length can *also* be written as $4x + 4$.

Therefore, $4 \cdot (x + 1) = 4x + 4$.

1. Write an expression for the repeated pattern in the model. Then multiply the expression using the distributive property.

a.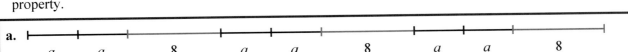

$3(2a + 8) =$

b.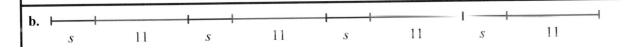

c.

2. Draw line segments to represent the expressions. Then, multiply the expressions using the distributive property.

a. $3(b + 8) =$

b. $4(2w + 1) =$

c. $2(3x + 5) =$

3. Use the distributive property to multiply.

a. $2(x + 9) =$	b. $7(4y + 5) =$	c. $10(9x + 8) =$
d. $8(x + y) =$	e. $s(4 + t) =$	f. $u(v + w) =$

4. The side of a regular hexagon is $2x + 4$. What is its perimeter?

5. The perimeter of a square is $24y + 40$. How long is its side?

$P = 24y + 40$

6. **a.** Write an expression for the total cost of buying n jars of coconut oil for $20 each.

b. What is the total cost, if an additional shipping cost of $11 is added to the order?

c. You repeat the same order three times during the year.
Multiply the expression from (b) by 3, and use the distributive property.

d. How many jars of coconut oil did you buy in a year, if you spent $393 in these three orders?

The distributive property works the same way *with subtraction* and with *more than two terms* in the second factor. The proof that it works with subtraction has to do with how negative numbers work in multiplication, and it is not presented here.

However, consider this numerical example. To multiply 7 · 98, think of the 98 as 100 − 2, and multiply in parts:
$$7(100 - 2) = 7 \cdot 100 - 7 \cdot 2 = 700 - 14 = 686$$

Example 3. $2(7x - y) = 2 \cdot 7x - 2 \cdot y = 14x - 2y$

Example 4. $8(s - t + 5) = 8s - 8t + 40$

7. Use the distributive property to multiply.

a. $11(x - 7) =$	b. $30(x + y + 5) =$	c. $10(r + 2s + 0.1) =$
d. $5(3x - 2y - 6) =$	e. $s(1.5 + t - x) =$	f. $0.5(3v + 2w - 7) =$

8. Solve mentally! (Hint: the distributive property will help.)

a. $8 \cdot 99 =$	b. $6 \cdot 98 =$	c. $5 \cdot 599 =$

The area of this *whole* rectangle is 9 times $(8 + b)$. But, if we think of it as *two* rectangles, the area of the first rectangle is 9 · 8, and of the second, 9 · b.

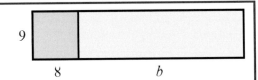

Of course, these two expressions are equal:
$$9 \cdot (8 + b) = 9 \cdot 8 + 9 \cdot b = 72 + 9b \text{ or } 9b + 72.$$

9. Write an expression for the area in two ways, thinking of the overall rectangle or its component rectangles.

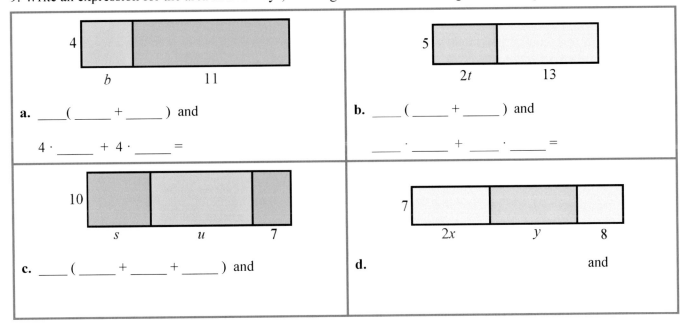

a. ____(____+____) and

4 · ____ + 4 · ____ =

b. ____(____+____) and

____ · ____ + ____ · ____ =

c. ____(____+____+____) and

d. and

37

10. Find the missing numbers or variables in these area models.

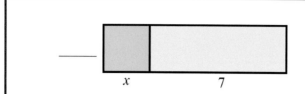

a. _____ (x + 7) = _____ x + 63

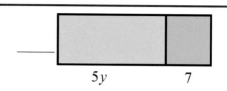

b. _____ (5y + 7) = 40y + _____

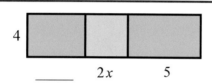

c. The total area is 12y + 8x + 20.

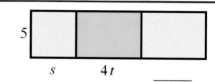

d. 5(s + 4t + _____) = 5s + 20t + 15v

11. Find the missing number or term in the equations.

a. _____ (20x + 3) = 200x + 30

b. 4(6s − _____) = 24s − 4x

c. 2(_____ + 1.5y + 0.9) = 7x + 3y + 1.8

d. 4(_____ − _____ + _____) = 0.4x − 1.2y + 1.6

12. Use the distributive property "backwards" to write the expression as a product. This is called **factoring**.

a. 2x + 6 = _____ (x + 3)

b. 4y + 16 = 4(_____ + _____)

c. 21t + 7 = _____ (3t + _____)

d. 16d + 24 = _____ (2d + _____)

e. 15x − 35 = _____ (_____ − _____)

f. 7a − 49 =

13. a. Sketch a rectangle with an area of 9x + 15.

b. Sketch a rectangle with an area of 9a + 15b + 3.

14. Factor these sums (write them as products). Think of divisibility!

a. 64x + 40 =

b. 54x + 18 =

c. 100y − 20 =

d. 90t + 33s + 30 =

e. 36x − 12y + 24 =

f. 2x + 8z − 40 =

The distributive property works with division, too. Just like we can multiply in parts, we can *divide in parts*.

Example 5. $\dfrac{50 + 35}{5}$ is the same as $\dfrac{50}{5} + \dfrac{35}{5}$. You can then do the two divisions separately and add last.

Example 6. $\dfrac{6x + y}{2}$ is the same as $\dfrac{6x}{2} + \dfrac{y}{2}$. And $\dfrac{6x}{2}$ simplifies to $3x$.

This works because any division can be rewritten as a multiplication by a fraction, and multiplication is distributive.

15. Divide in parts using mental math. You may end up with a fraction in the answer.

a. $\dfrac{300 + 2}{3}$	b. $\dfrac{13 + 700}{7}$	c. $\dfrac{5{,}031}{5}$
d. $\dfrac{5x - 3}{6}$	e. $\dfrac{x + 7}{7}$	f. $\dfrac{4x + 2}{4}$

16. The Larson family are planning their new house. It is going to be 25 ft on one side and have a garage that is 15 ft wide, but they have not decided on the length of the house yet.

    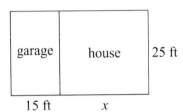

    a. If the total area of the house + garage is limited to 1200 square feet, how long can the house be?

    b. Write a single equation for the question above. Write it in the form "(formula for area) = 1200." You do not have to solve the equation—just write it.

We can even model expressions with subtraction, such as $3(7 - 2)$, using an area model. We use dark shading to indicate that an area is subtracted ("negative" area).

Puzzle Corner

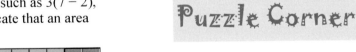

For example, the picture above illustrates $3(7 - 2) = 3 \cdot 7 - 3 \cdot 2$.

a. What expression is modeled below? 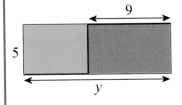	b. Draw a model for $3(x + y - 2) = 3x + 3y - 6$.

Review 1

1. Find the value of these expressions.

| a. $(6+4)^2 \cdot (12-9)^3$ | b. $3 \cdot (5-(7-5))$ | c. $\dfrac{(5-3) \cdot 2}{2^3} + 7$ |

2. Name the property of arithmetic illustrated by the fact that $(5 \cdot z) \cdot 3$ is equal to $5 \cdot (z \cdot 3)$.

3. Evaluate the expressions.

| a. $100 - 2x^2$, when $x = 5$ | b. $\dfrac{2s}{s^3 + 3}$, when $s = 4$ |

4. Which equation matches the situation? *Hint: Give the variable(s) some value(s) to test the situation.*

   a. The shorter beam (length l_1) is 1.5 meters shorter than the longer beam (length l_2).

   $l_1 = 1.5 - l_2$ $l_2 = 1.5 - l_1$ $l_2 = l_1 - 1.5$ $l_1 = l_2 - 1.5$

   b. The dog lost 1/6 of its original weight (w), and weighs now 23 kg.

   $\dfrac{w}{6} = 23$ $\dfrac{5w}{6} = 23$ $\dfrac{6w}{5} = 23$ $w - 1/6 = 23$ $w - 5/6 = 23$

5. Is subtraction commutative? In other words, is it true that $a - b$ has the same value as $b - a$, no matter what values we use for a and b? Explain your reasoning.

6. Write a SINGLE expression to match these situations.

   a. A pair of jeans costs p dollars. The jeans are now discounted by 1/5 of that price. Write an expression for the discounted price.

   b. It costs Mandy $0.18 to drive her car one mile. How much does it cost her to drive y miles? Write an expression.

   c. The pet store sells dog food in bags of two different sizes: 3-kg and 8-kg. What is the total weight of n of the smaller bags and m of the larger bags?

7. Simplify the expressions.

a. $x + 2 + x + x$	b. $x \cdot 2 \cdot x \cdot x \cdot x$	c. $8v + 12v$
d. $8v \cdot 12v$	e. $4z \cdot 9z \cdot z$	f. $f + 2f + 10g - f - 4g$

8. **a.** Sketch a rectangle that is $5x$ tall and $2x$ wide.

   b. What is its area?

   c. What is its perimeter?

9. Use the distributive property to multiply.

a. $12(v - 9)$	b. $3(a + b + 2)$	c. $3(0.5t - x)$

10. Draw a diagram of two rectangles to illustrate that the product $11(x + 7)$ is equal to $11x + 77$.

11. Fill in the table.

Expression	the terms in it	coefficient(s)	Constants
a^8			
$2x + 9y$			

12. The perimeter of a regular pentagon is $30s + 45$. How long is one side?

13. Factor these sums (write them as products). Think of divisibility!

a. $48x + 12 =$	b. $40x - 25 =$
c. $6y - 2z =$	d. $56t - 16s + 8 =$

Solving Equations

Do you remember? An **equation** has two expressions, separated by an equal sign:

(expression) = (expression)

To solve an equation, we can

- add the same quantity to both sides
- subtract the same quantity from both sides
- multiply both sides by the same number
- divide both sides by the same number

Notice that in any of these operations, the two expressions on the left and right sides of the equation will remain equal, even though the expressions themselves change!

Example 1. We will manipulate the simple equation $2 + 3 = 5$ in these four ways. We will write in the margin the operation that is going to be done next to both sides.

Let's add six to both sides.	$2 + 3 = 5 \quad \vert + 6$
Now, both sides equal 11. Next, we multiply both sides by 8.	$2 + 3 + 6 = 11 \quad \vert \cdot 8$
Now, both sides equal 88. Next, we subtract 12 from both sides.	$16 + 24 + 48 = 88 \quad \vert - 12$
Now both sides equal 76. Next, we divide both sides by 2.	$16 + 24 + 48 - 12 = 76 \quad \vert \div 2$
Now both sides equal 38.	$8 + 12 + 24 - 6 = 38$

Of course, you do not usually work with equations like the one above, but with ones that have an unknown. Your goal is to **isolate** the unknown, or **leave it by itself,** on one side. Then the equation is solved.

We can model an equation with a **pan balance**. Both sides (pans) of the balance will have an *equal* weight in them, thus the sides are balanced (not tipped to either side).

Example 2. Solve the equation $x - 2 = 3$.

We can write this equation as $x + (-2) = 3$ and model it using negative and positive counters in the balance.

Here x is accompanied by two negatives on the left side. Adding two <u>positives</u> *to both sides* will cancel those two negatives. We denote that by writing "**+2**" in the margin.

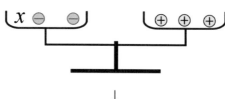

margin ↓

$x + (-2) = 3 \quad \vert + 2$

We write $x + (-2) + 2 = 3 + 2$ to show that 2 was added to both sides of the equation.

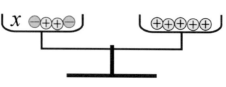

$x + (-2) \,\boxed{+ 2} = 3 \,\boxed{+ 2}$

Now the two positives and two negatives on the left side cancel each other, and x is left by itself. On the right side we have 5, so x equals 5 positives.

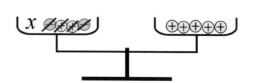

$x = 5$

1. Solve the equations. Write in the margin what operation you do to both sides.

a. Balance	Equation	Operation to do to both sides
	$x + 1 = -4$	

b. Balance	Equation	Operation to do to both sides
	$x - 1 = -3$	

c. Balance	Equation	Operation
	$x - 2 = 6$	

d. Balance	Equation	Operation
	$x + 5 = 2$	

2. If you need more practice, solve the following equations also. Draw a balance in your notebook to help you.

    a. $x + (-3) = 7$ **b.** $x + (-3) = -4$ **c.** $x + 6 = -1$ **d.** $x + 5 = -4$

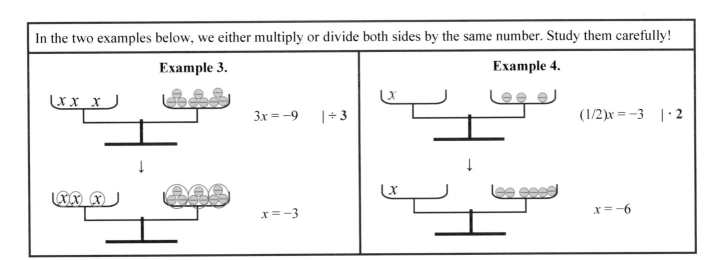

In the two examples below, we either multiply or divide both sides by the same number. Study them carefully!

Example 3. $3x = -9 \quad | \div 3$
$x = -3$

Example 4. $(1/2)x = -3 \quad | \cdot 2$
$x = -6$

3. Solve the equations. Write in the margin what operation you do to both sides.

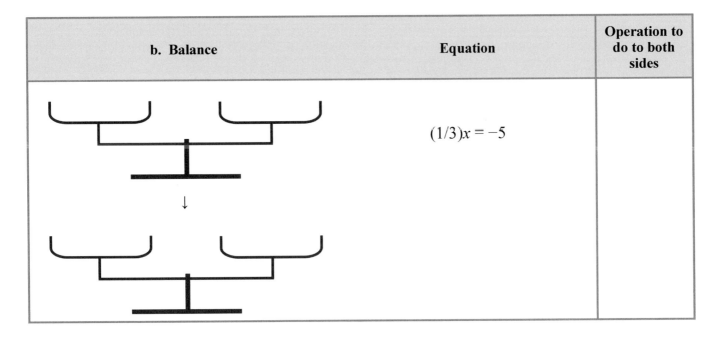

a. Balance	Equation	Operation to do to both sides
	$4x = -12$	

b. Balance	Equation	Operation to do to both sides
	$(1/3)x = -5$	

4. Let's review a little! Which equation matches the situation?

   a. A stuffed lion costs $8 less than a stuffed elephant. Note: p_l signifies the price of the lion and p_e the price of the elephant.

   | $p_e = 8 - p_l$ | $p_e = p_l - 8$ | $p_l = p_e - 8$ |

   b. A shirt is discounted by 1/5, and now it costs $16.

   | $p - 1/5 = \$16$ | $\dfrac{4p}{5} = \$16$ | $\dfrac{p}{5} = \$16$ | $\dfrac{5p}{4} = \$16$ | $\dfrac{p}{5} = 4 \cdot \$16$ |

5. Find the roots of the equation $\dfrac{6}{x+1} = -3$ in the set $\{2, -2, 3, -3, 4, -4\}$.

6. Write an equation, then solve it using guess and check. Each root is between −20 and 20.

a. 7 less than x equals 5.
b. 5 minus 8 equals x plus 1
c. The quantity x minus 1 divided by 2 is equal to 4.
d. x cubed equals 8
e. −3 is equal to the quotient of 15 and y
f. Five times the quantity x plus 1 equals 10.

Example 5. Solve $2x + 5 = -3$.
The solution requires two steps.

The two x's are accompanied by five positives. Therefore, we will subtract five *from both sides*.

Subtracting 5 is the same as adding -5, so the right side ends up with 8 negatives.

The positives and negatives on the left side cancel each other, and $2x$ is left by itself on that side.

Now we need to divide both sides by 2. Again, we note that in the margin.

We can see that x equals 4 negatives.

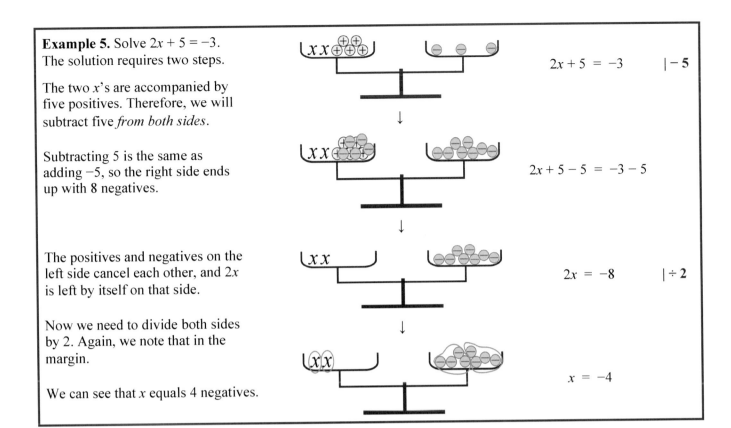

$2x + 5 = -3 \quad | -5$

$2x + 5 - 5 = -3 - 5$

$2x = -8 \quad | \div 2$

$x = -4$

7. Solve the equations. Write in the margin what operation you do to both sides.

a. Balance	Equation	Operation to do to both sides
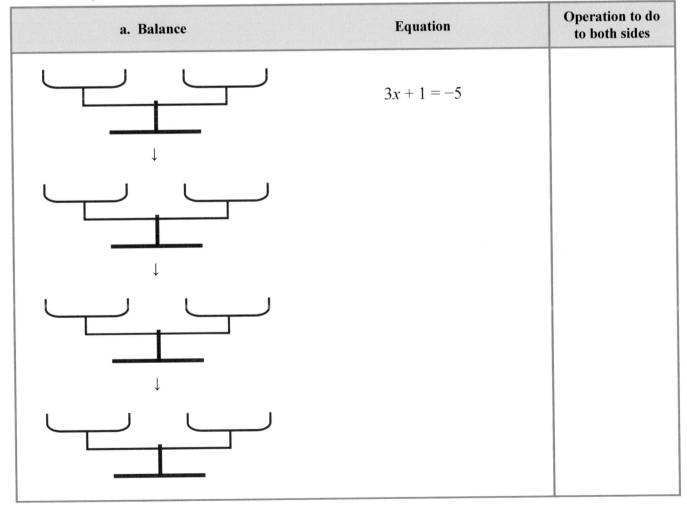	$3x + 1 = -5$	

47

b. Balance	Equation	Operation
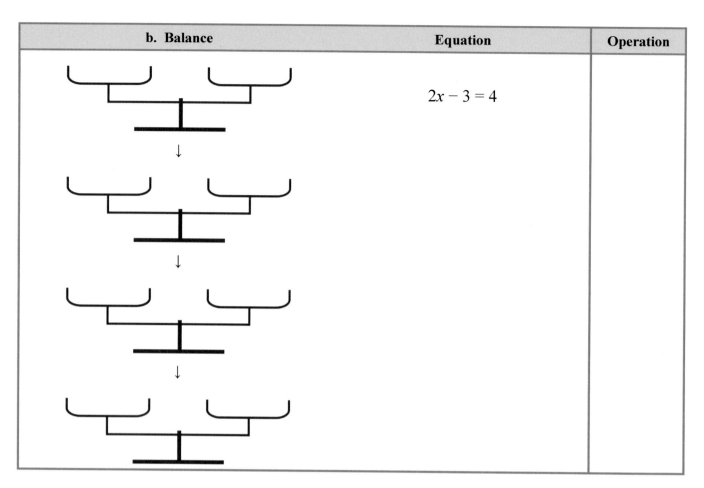	$2x - 3 = 4$	

c. Balance	Equation	Operation
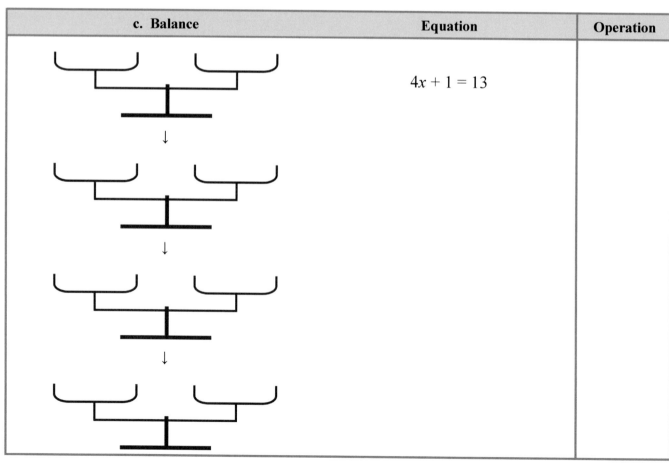	$4x + 1 = 13$	

Addition and Subtraction Equations

You can **keep track of the operations you're using in a couple of different ways**. One way is to write the operation underneath the equation on both sides. Another is to write it in the right margin, like we did in the last lesson. But in either case, always check your solution: does it solve the original equation?	**One way:** $x + 9 = 4$ $\underline{-9 \quad -9}$ $x = -5$	**Another way:** $x + 9 = 4 \quad \mid -9$ $x + 9 - 9 = 4 - 9$ (This step is optional.) $x = -5$
	Check: $-5 + 9 \stackrel{?}{=} 4$. Yes, it checks.	

1. Solve these one-step equations. Keep track of the operations either under the equation or in the margin, whichever way your teacher prefers.

a. $\quad x + 5 = 9$	b. $\quad x + 5 = -9$
c. $\quad x - 2 = 3$	d. $\quad w - 2 = 3$
e. $\quad z + 5 = 0$	f. $\quad y - 8 = -7$

2. In these equations, your first step is to simplify what is on the right side.

a. $\quad x - 7 = 2 + 8$ $\quad x - 7 = 10$	b. $\quad x - 10 = -9 + 5$
c. $\quad s + 5 = 3 + (-9)$	d. $\quad t + 6 = -3 - 5$

If the **unknown is on the right side of the equation**, you have two options:

- First, flip the two sides. Then solve as usual.

- Or, solve as usual, isolating the unknown —this time on the right side of the equation. The solution will initially read as $-7 = x$. Flip the sides now and write the solution as $x = -7$.

1. Flip the sides first:

$$-9 = x - 2$$
$$x - 2 = -9$$
$$\underline{+2 \qquad +2}$$
$$x \quad = -7$$

2. Solve as it is:

$$-9 = x - 2$$
$$\underline{+2 \qquad +2}$$
$$-7 = x$$
$$x = -7$$

3. Solve. Check your solutions.

a.	$-8 = s + 6$	b.	$-2 = x - 7$	
c.	$4 = s + (-5)$	d.	$2 - 8 = y + 6$	
e.	$5 + x = -9$	f.	$-6 - 5 = 1 + z$	
g.	$y - (-7) = 1 - (-5)$	h.	$6 + (-2) = x - 2$	
i.	$3 - (-9) = x + 5$	j.	$2 - 8 = 2 + w$	

Example 1. Solve $-2 + 8 = -x$.

Our first step is to simplify the sum $-2 + 8$.

The equation becomes $-x = 6$ (or $6 = -x$). What does that mean? It means that the <u>opposite of x</u> is 6. So x must equal -6!

Lastly we check the solution $x = -6$:
$-2 + 8 \stackrel{?}{=} -(-6)$, which simplifies to $6 \stackrel{?}{=} 6$, so it checks.

1. Flip the sides first:	2. Solve as it is:
$-2 + 8 = -x$	$-2 + 8 = -x$
$-x = -2 + 8$	$6 = -x$
$-x = 6$	$-6 = x$
$x = -6$	$x = -6$

4. Solve for x. Check your solutions.

a. $-x = 6$	b. $-x = 5 - 9$
c. $4 + 3 = -y$	d. $-2 - 6 = -z$

5. Which equation best matches the situation?

   a. The sides of a square playground were shortened by 1/2 m, and now its perimeter is 12 m.

   $4s - 1/2 = 12$ $4(s - 1/2) = 12$ $4s - 50 = 12$

   $s - 1/2 = 4 \cdot 12$ $s - 1/2 = 12$ $4(s - 0.5) = 12$

   b. How long were the sides before they were made shorter? Solve the problem using mental math.

   c. *Challenge*: Solve the same problem using the equation. Compare the steps of this formal solution to the way you reasoned it out in your head. Are the steps similar?

6. Here is another "growing pattern." Draw steps 4 and 5 and answer the questions.

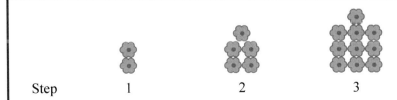

Step   1   2   3

a. How do you see this pattern grow?

b. How many flowers will there be in step 39?

c. In step n?

Example 2. Solve $8 - x = -2$.

As usual, think about what you need to do to isolate the x on one side. Since there is an 8 on the side with the x, we need to subtract 8 from both sides.

However, note that x is being *subtracted*, or in other words, there is a negative sign in front of the x.

This negative sign <u>does not disappear</u> when you subtract 8 from both sides.

Writing the operation underneath each side:

$$
\begin{array}{rcr}
8 - x & = & -2 \\
-8 & & -8 \\
\hline
-x & = & -10 \\
x & = & 10
\end{array}
$$

Writing the operation in the right margin:

$$
\begin{array}{rcr}
8 - x & = & -2 \quad |-8 \\
8 - x - 8 & = & -2 - 8 \\
-x & = & -10 \\
x & = & 10
\end{array}
$$

If this is confusing, think of it this way: The equation $8 - x = -2$ can also be written as $8 + (-x) = -2$. When we subtract 8 from both sides, the left side becomes $8 + (-x) - 8$. The positive 8 and negative 8 will cancel each other and leave $-x$.

So we end up with the equation $-x = -10$. This equation says that the opposite of x is negative 10, so x must be 10. (Why?)

Lastly, check your solution by substituting $x = 10$ back into the original equation: $8 - \underline{10} \overset{?}{=} -2$ ✓

7. Solve. Check your solutions.

a.	$2 - x = 6$		**b.**	$8 - x = 7$
c.	$-5 - x = 5$		**d.**	$2 - x = -6$
e.	$1 = -5 - x$		**f.**	$2 + (-9) = 8 - z$
g.	$-8 + r = -5 + (-7)$		**h.**	$2 - (-5) = 2 + 5 + t$

Multiplication and Division Equations

Do you remember **how to show simplification**? Just cross out the numbers and write the new numerator above the fraction and the new denominator below it. Notice that the number you divide by (the 5 in the fraction at the right) isn't indicated in any way!	$\dfrac{\overset{7}{\cancel{35}}}{\underset{11}{\cancel{55}}} = \dfrac{7}{11}$
We can simplify expressions involving variables in exactly the same way. In the examples on the right, we cross out the *same number* from the numerator and the denominator. That is based on the fact that a number divided by itself is 1. We could write a little "1" beside each number that is crossed out, but that is usually omitted.	$\dfrac{\cancel{2}x}{\cancel{2}} = x \qquad \dfrac{\cancel{5}s}{\cancel{5}} = s$ $\dfrac{4\cancel{x}}{\cancel{x}} = 4$
In this example, we simplify the fraction 3/6 into 1/2 the usual way.	$\dfrac{\overset{1}{\cancel{3}}x}{\underset{2}{\cancel{6}}} = \dfrac{1}{2}x \text{ or } \dfrac{x}{2}$
Notice: We divide both the numerator and the denominator by 8, but <u>this leaves −1 in the denominator</u>. Therefore, the whole expression simplifies to −z instead of z.	$\dfrac{\cancel{8}z}{\cancel{-8}} = \dfrac{z}{-1} = -z$

1. Simplify.

a. $\dfrac{8x}{8}$	b. $\dfrac{8x}{2}$	c. $\dfrac{2x}{8}$
d. $\dfrac{-6x}{-6}$	e. $\dfrac{-6x}{6}$	f. $\dfrac{6x}{-6}$
g. $\dfrac{6w}{2}$	h. $\dfrac{6w}{w}$	i. $\dfrac{6w}{-2}$

2. Draw the fourth and fifth steps of the pattern and answer the questions.

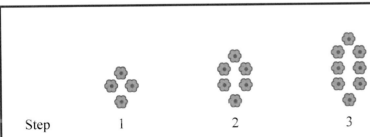

Step   1   2   3

a. How would you describe the growth of this pattern?

b. How many flowers will there be in step 39?

c. In step *n*?

Now you should be ready to use multiplication and division to solve simple equations.

Example 1. Solve $-2x = 68$.

The unknown is being multiplied by -2. To isolate it, we need to divide both sides by -2. (See the solution on the right.)

We get $x = -34$. Lastly we check the solution by substituting -34 in the place of x in the original equation:

$-2(-34) \stackrel{?}{=} 68$

$68 = 68$ It checks.

$-2x = 68$ This is the original equation.

$\dfrac{-2x}{-2} = \dfrac{68}{-2}$ We divide both sides by -2.

$\dfrac{\cancel{-2}x}{\cancel{-2}} = \dfrac{68}{-2}$ Now it is time to simplify. We cross out the -2 factors on the left side. On the right side, we do the division.

$x = -34$ This is the final answer.

Note: Most people combine the first 3 steps into one when writing the solution. Here they are written out for clarity.

3. Solve. Check your solutions.

a.	$5x = -45$		**b.**	$-3y = -21$
c.	$-4 = 4s$		**d.**	$72 = -6y$

4. Solve. Simplify the one side first.

a.	$-5q = -40 - 5$		**b.**	$2 \cdot 36 = -6y$
c.	$3x = -4 + 3 + (-2)$		**d.**	$5 \cdot (-4) = -10z$

Example 2. Solve $\frac{x}{-6} = -5$.

Here the unknown is divided by −6. To undo that division, we need to *multiply* both sides by −6. (See the solution on the right.)

We get $x = 30$. Lastly we check the solution:

$$\frac{30}{-6} \stackrel{?}{=} -5$$

$$-5 = -5 \checkmark$$

$\frac{x}{-6} = -5$		This is the original equation.
$\frac{x}{-6} \cdot (-6) = -5 \cdot (-6)$		We multiply both sides by −6.
$\frac{x}{\cancel{-6}} \cdot (\cancel{-6}) = 30$		Now it is time to simplify. We cross out the −6 factors on the left side, and multiply on the right.
$x = 30$		This is the final answer.

When writing the solution, most people would combine steps 2 and 3. Here both are written out for clarity.

5. Solve. Check your solutions.

a. $\quad \frac{x}{2} = -45$

b. $\quad \frac{s}{-7} = -11$

c. $\quad \frac{c}{-7} = 4$

d. $\quad \frac{a}{-13} = -9 + (-11)$

6. Write an equation for each situation. Then solve it. Do not write the answer only, as the main purpose of this exercise is to practice writing equations.

   a. A submarine was located at a depth of 500 ft.
      There was a shark swimming at 1/6 of that depth.
      At what depth is the shark?

   b. Three towns divided highway repair costs equally.
      Each town ended up paying $21,200.
      How much did the repairs cost in total?

Example 3. Solve $-\frac{1}{5}x = 2$. Here the unknown is multiplied by a negative fraction, but do not panic!

You see, you can *also* write this equation as $\frac{x}{-5} = 2$, where the unknown is simply divided by negative 5.

So what should we do in order to isolate *x*?

That is correct! Multiplying by −5 will isolate *x*. In the boxes below, this equation is solved in two slightly different ways, though both are doing essentially the same thing: multiplying both sides by −5.

Multiplying a fraction by its reciprocal:	**Canceling a common factor:**
$-\frac{1}{5}x = 2 \quad \vert \cdot (-5)$	$-\frac{1}{5}x = 2 \quad$ rewrite the equation
$(-5) \cdot \left(-\frac{1}{5}\right)x = (-5) \cdot 2 \quad$ Note that −5 times −1/5 is 1.	$\frac{x}{-5} = 2 \quad \vert \cdot (-5)$
$1x = -10$	$\frac{x}{\cancel{-5}} \cdot (\cancel{-5}) = 2 \cdot (-5)$
$x = -10$	$x = -10$

Lastly we check the solution by substituting −10 in place of *x* in the original equation:

$-\frac{1}{5}(-10) \stackrel{?}{=} 2$

$2 = 2$ It checks.

7. Solve. Check your solutions.

a. $\frac{1}{3}x = -15$	**b.** $-\frac{1}{6}x = -20$	**c.** $-\frac{1}{4}x = 18$
d. $-2 = -\frac{1}{9}x$	**e.** $-21 = \frac{1}{8}x$	**f.** $\frac{1}{12}x = -7 + 5$

Word Problems

Example 1. The area of a rectangle is 195 m². One side measures 13 m. How long is the other side?

To write the equation, you need to remember that the area of any rectangle is calculated as **area = side · side**.

That relationship gives us our equation. We simply substitute the known area and the length of the known side into the equation and represent the length of the unknown side by some variable:

$$195 = s \cdot 13$$

Then we rewrite the expression $s \cdot 13$ in the usual way, where the coefficient (13) comes first and the variable(s) last. We also flip the sides of the equation, so the unknown is on the left: $13s = 195$.

Solution:

$$13s = 195$$
$$\frac{13s}{13} = \frac{195}{13} \quad \text{We divide both sides by 13 and simplify.}$$
$$s = 15 \quad \text{This is the solution.}$$

Check:

$$13 \cdot 15 \stackrel{?}{=} 195$$
$$195 = 195 \checkmark$$

For each given situation, write an equation and solve it. The problems themselves are simple, and you could solve them without writing an equation, but it is important to practice writing equations! You need to learn to write equations for simple situations now, so you will be able to write equations for more complex situations later on.

1. The perimeter of a square is 456 cm. How long is one side?

   Equation:

2. The area of a rectangular park is 4,588 square feet. One side measures 62 feet. How long is the other side?

   Equation:

3. John bought some boxes of screws for $15 each and paid a total of $165. How many boxes did he buy?

   Equation:

4. A candle burned at a steady rate of 2 cm per hour for 4 hours. Now it is only 6 cm long. How long was the candle at first?

   Equation:

5. A baby dolphin is only 1/12 as heavy as his mother. The baby weighs 15 kilograms. How much does the mother dolphin weigh?

   Equation:

Example 2. Solve the equation $45 + y + 82 + 192 = 374$ using a bar model and also using an equation.

Bar model:

For the bar model, we draw the addends as parts of the total, which we indicate with a double-headed arrow.

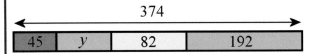

To solve it, we subtract all the known parts from 374:

$y = 374 - 45 - 82 - 192$

$y = 55$

Equation:

$45 + y + 82 + 192 = 374$

$319 + y = 374$

$319 + y - 319 = 374 - 319$

$y = 55$

First, add what you can on the left side.

Next, subtract 319 from both sides.

Check: $45 + 55 + 82 + 192 \stackrel{?}{=} 374$

$374 = 374$ ✓

6. Using both a bar model and an equation, solve the equation $21 + x + 193 = 432$.

Bar model:	Equation:

7. Using both a bar model and an equation, solve the equation $495 + 304 + w + 94 = 1{,}093$.

Bar model:	Equation:

Constant Speed

If an object travels with a constant speed, we have three quantities to consider: *speed or velocity* (*v*), *time* (*t*), and *distance* (*d*). The formula $d = vt$ tells us how they are interconnected.

Does that formula make sense?

Let's say John rides his bicycle at a constant speed of 12 km per hour for four hours. How far can he go? The formula says you multiply the speed (12 km/h) by the time (4 h) to get the distance (48 km). So the formula does make sense — that is how our "common sense" tells us to calculate it also.

Example 1. A boat travels at a constant speed of 15 km/h. How long will it take the boat to go a distance of 21 km?

The problem gives us the speed and the distance. The time (*t*) is unknown.

We can solve the unknown time by using the formula $d = vt$. We simply substitute the given values of *v* and *d* in it, and we will get an equation that we can solve:

$$d = v\ t$$
$$\downarrow \quad \downarrow$$
$$21 = 15\ t$$

To make it easier, we will leave off the units while solving the equation. We can do that, since both the velocity and the distance involve kilometers.

Next, we simply solve this equation:

$21 = 15t$		Flip the sides.
$15t = 21$		Divide both sides by 15.
$\dfrac{15t}{15} = \dfrac{21}{15}$		The 15 in numerator and denominator cancel.
$t = 1\ 6/15$		

The final answer $t = 1\ 6/15$ is in *hours*, because the unit for speed was kilometers per <u>hour</u>.

Let's work on that answer a bit more. First, it simplifies to 1 2/5 hours. Then let's change 2/5 of an hour to minutes.

How much is 1/5 of an hour? That's right, it is 12 minutes. And 2/5 of an hour is 24 minutes. So the final answer is 1 hour, 24 minutes.

You may use a calculator for all the problems in this lesson.

1. Use the formula $d = vt$ to solve the problems.

a. A caterpillar crawls along at a constant speed of 20 cm/min. How long will it take it to travel 34 cm?

$$d = v\ t$$
$$\downarrow \quad \downarrow \quad \downarrow$$

b. Father leaves at 7:40 a.m. to drive 20 miles to work. If his average speed is 48 mph, when will he arrive at work?

$$d = v\ t$$
$$\downarrow \quad \downarrow \quad \downarrow$$

How to change hours and minutes into fractional and decimal hours and vice versa	
Example 2. Change 14 minutes into hours. Since there are 60 minutes in an hour, 14 minutes is simply 14/60 of an hour. This simplifies to 7/30 hours. You can also change it into a decimal by dividing to get 0.233 hours (rounded to three decimals).	**Example 3.** Change 4.593 hours into hours and minutes. How many minutes are in the decimal part? Since one hour is 60 minutes, 0.593 hours is 0.593 · 60 minutes = 35.58 minutes ≈ 36 minutes. So 4.593 hours ≈ 4 hours 36 minutes.

2. Convert the given times into hours in decimal format. Round your answers to three decimal digits.

a. 35 minutes	**b.** 44 minutes
c. 2 h 16 min	**d.** 4 h 9 min

3. Give these times in hours and minutes.

a. 2.4 hours	**b.** 0.472 hours
c. 3 3/5 hours	**d.** 16/50 hours

4. The average speed of a bus is 64 km/hour. What distance can it travel in 4 hours and 15 minutes?

5. Sam is an athlete who can run 10 miles in an hour. How long will it take him to run home from the shopping center, a distance of 2.4 miles?

6. A train traveled a distance of 360 miles between two towns so that in the first half of the distance, its average speed was 90 mph, and in the second half, only 75 mph. How long did it take to travel from the one town to the other?

7. Elijah wants to use the extra time between classes for exercising. He plans to jog for 25 minutes in one direction, turn, and jog back to school. What is the distance Elijah can jog in 25 minutes if his average jogging speed is 6 mph?

Example 4. Andy drove from his home to his workplace, which was 22 miles away, in 26 minutes. What was his average speed?

The average speed of cars is usually given in miles per hour or kilometers per hour. This unit of speed actually gives us a **formula for calculating speed**:

$$\begin{array}{ccccc} \text{speed} & \text{is} & \text{miles} & \text{per} & \text{hour} \\ \downarrow & \downarrow & \downarrow & \downarrow & \downarrow \\ \text{velocity} & = & \text{distance} & / & \text{time} \end{array}$$

(The word "per" indicates division.)

In symbols, $v = \dfrac{d}{t}$.

His average speed is therefore

$$v = \frac{d}{t} = \frac{22 \text{ miles}}{26 \text{ minutes}} = \frac{11}{13} \text{ miles per minute}$$

≈ 0.846 miles per minute.

The problem is that the average speed is usually given in miles per *hour*, not miles per minute. How can we fix that?

One way is to multiply our answer by 60. Doing that, we get 0.846 mi/min · 60 min/hr = 50.76 ≈ 51 mph.

Another way is to change the original time of 26 minutes into hours before using the formula.

Now 26 minutes is simply 26/60 hours, which simplifies to 13/30 hours. We can write

$$v = \frac{22 \text{ miles}}{13/30 \text{ hours}}$$

This is a **complex fraction**: a fraction that has another fraction in the numerator, denominator, or both.

One way to calculate its value with a calculator is to use parentheses and input it as 22 ÷ (13 ÷ 30). Check out what happens if you input it as 22 ÷ 13 ÷ 30.

Tip: Instead of parentheses, you can use the reciprocal button (1/x) on your calculator. First calculate the value of the fraction inverted (upside-down): $\dfrac{13/30}{22}$. This is the **reciprocal** of the fraction. You can input it as 13 ÷ 30 ÷ 22.

Once you've calculated the reciprocal, push the 1/x button to convert the reciprocal into the answer that you want.

---

8. Find the average speed in the given units.

   a. A duck flies 3 miles in 6 minutes.
   Give your answer in miles per hour.

   b. A lion runs 900 meters in 1 minute.
   Give your answer in kilometers per hour.

   c. Henry sleds 75 meters down the hill in 1.5 minutes.
   Give your answer in meters per second.

   d. Rachel swims 400 meters in 32 minutes.
   Give your answer in kilometers per hour.

9. Jake's grandparents live 150 km away from his home. One day it took him 2 h 14 min to get there and 1 h 55 min to come back home. In the questions below, round your answers to one decimal digit.

   a. What was his average speed going there?
   Hint: Change the time in hours and minutes into decimal hours.

   b. What was his average speed coming back?

   c. What was his overall average speed for the whole trip?

10. Another day Jake visited his grandparents again. Let's say that, because of the traffic, Jake achieved an average speed of 75 km/h going there but an average speed of only 65 km/h coming back. How much longer did Jake spend driving home from his grandparents' place than going there?

How to remember the formula $d = vt$	Just solve for d from the formula $v = d/t$
I will show you how to *derive* that formula! Then you do not really have to memorize it. But you *do* need to remember the formula $v = d/t$. You can remember *that* formula with the trick I explained earlier—by thinking of the common unit for measuring speed (miles per hour): speed is miles per hour ↓ ↓ ↓ ↓ ↓ velocity = distance / time In symbols: $v = d/t$.	$v = \dfrac{d}{t}$ We want d alone, so we multiply both sides by t. $vt = \dfrac{d}{t} \cdot t$ The t's on the right side cancel. $vt = d$ We have it! Turning it around we get $d = vt$, which is the most common formula to show how velocity (v), time (t), and distance (d) are related.

11. Compare the average speeds to find which bird is faster: a seagull that flies 10 miles in 24 minutes or an eagle that flies 14 miles in half an hour?

12. A train normally travels at a speed of 120 km per hour. One day, the conditions were so icy and cold that it had to slow down to travel safely. So the train traveled the first half of its 160-km journey at half its normal speed. Then the weather improved, and the train was able to go faster again. It sped the remaining distance at twice its normal speed to make up time.

   a. How long did the train take to travel the whole distance (160 km)?

   b. How long would the train have taken if it had traveled the whole trip at its normal speed?

   c. What was the train's average speed for the trip on this cold and icy day?

13. How long will it take Charlotte to ride her bike from the music store to her home — a distance of 4.5 km — if she rides 1/3 of it at 12 km/h and the rest at 15 km/h?

The next problems are more challenging.

14. Your normal walking speed is 6 miles per hour. One day you walk slowly, at 3 miles per hour, half the distance from home to the swimming pool. Can you now make up for your slow walking by walking the remaining distance at double your normal speed?

    (Hint: Make up a distance between your home and the swimming pool for an example calculation. Choose an easy number.)

15. An airplane normally flies at a speed of 1,000 km/h. Due to some turbulence, it has to travel at a lower speed of 800 km/h for the first 40 minutes of a 1600-km trip. How fast should it fly for the rest of the trip so as to make up for the lost time?

    (Hint: You will also need to calculate the normal traveling time for this trip.)

Two-Step Equations

Just like the name says, **two-step equations take two steps to solve.** We need to apply two different operations to both sides of the equation. Study the examples carefully. It is not difficult at all!

Example 1. Solve $2x + 3 = -5$.

On the side of the unknown (left), there is a multiplication by 2 and an addition of 3. To isolate the unknown, we need to undo those operations.

$$\begin{aligned} 2x + 3 &= -5 \quad &|-3 \\ 2x &= -8 \quad &|\div 2 \\ x &= -4 \end{aligned}$$

Check:
$$\begin{aligned} 2 \cdot (-4) + 3 &\stackrel{?}{=} -5 \\ -8 + 3 &\stackrel{?}{=} -5 \\ -5 &= -5 \checkmark \end{aligned}$$

What if you divide first?

In this equation you *could* start by dividing by 2 and then subtract next. However, it is easier to subtract first, then divide, because that way you avoid dealing with fractions.

The solution below shows the steps if you divide by 2 first. Notice that the 3 on the left side also has to be divided by 2 to become 3/2.

$$\begin{aligned} 2x + 3 &= -5 \quad &|\div 2 \\ x + (3/2) &= -5/2 \quad &|-3/2 \\ x &= -5/2 - 3/2 \\ x &= -4 \end{aligned}$$

1. Solve. Check your solutions (as always!).

a.	$5x + 2 = 67$	**b.**	$3y - 2 = 71$
c.	$-2x + 11 = 75$	**d.**	$8z - 2 = -98$

Example 2. In the equation below, the easiest thing to do is to multiply by 7 first, and then subtract.

$$\frac{x+2}{7} = 12 \quad | \cdot 7$$

$$\frac{7 \cdot (x+2)}{7} = 84$$

$$x + 2 = 84 \quad | -2$$

$$x = 82$$

Check:

$$\frac{82+2}{7} \stackrel{?}{=} 12$$

$$\frac{84}{7} \stackrel{?}{=} 12$$

$$12 = 12 \checkmark$$

2. Solve. Check your solutions (as always!).

a. $\dfrac{x+6}{5} = 14$

b. $\dfrac{x+2}{7} = -1$

c. $\dfrac{x-4}{12} = -3$

d. $\dfrac{x+1}{-5} = -21$

What if, in example 2, you subtract 2 first? (Optional)

It does not work. Subtracting 2 from both sides makes the equation more complicated.

$$\frac{x+2}{7} = 12 \quad | -2$$

$$\frac{x+2}{7} - 2 = 10$$

The quantity $(x + 2)$ is <u>also divided by 7.</u> It is $(x + 2)/7$. Notice that, when calculating the value of this expression, the parentheses indicate that the addition is to be done first, and the division last. Therefore, when *undoing* the operations, we need to undo the division first.

However, it is possible to subtract first. But you need to subtract 2/7 instead of 2. To see that, let's write the expression on the left side in a different way:

$$\frac{x+2}{7} = 12$$

$$\frac{x}{7} + \frac{2}{7} = 12 \quad | -2/7$$

$$\frac{x}{7} = 11\ 5/7 \quad | \cdot 7$$

$$x = 82$$

Because it involves fractions, this solution is more complicated than the one shown in Example 2.

Example 3. Again, the unknown is "tangled up" with two different operations (division and addition), so to isolate it, we need two steps.

$$\frac{x}{4} + 5 = -2 \quad \Big| -5$$

$$\frac{x}{4} = -7 \quad \Big| \cdot 4$$

$$x = -28$$

Check:
$$\frac{-28}{4} + 5 \stackrel{?}{=} -2$$

$$-7 + 5 \stackrel{?}{=} -2$$

$$-2 = -2 \checkmark$$

Solving it another way (optional)

In case you wonder if we could multiply by 4 first, yes, in this case we can.

$$\frac{x}{4} + 5 = -2 \quad \Big| \cdot 4$$

$$x + 20 = -8 \quad \Big| -20$$

$$x = -28$$

Note: In the first step, <u>both</u> terms on the left side (x/4 and 5) have to be multiplied by 4!

It is a common student error to multiply only the first term of an expression by a number and to forget to multiply the other terms by that number.

3. Solve. Compare equation (a) to equation (b). They are similar, yet different! Make sure you know how to solve each one.

a. $\frac{x}{10} + 3 = -2$

b. $\frac{x+3}{10} = -2$

4. Solve. Compare equation (a) to equation (b). They are similar, yet different! Make sure you know how to solve each one.

a. $\frac{x}{7} - 8 = -5$

b. $\frac{x-8}{7} = -5$

Example 4. What's different about this one? Check:

$$6 - 2n = 9 \quad |-6$$
$$-2n = 3 \quad |\div(-2)$$
$$n = -1\ 1/2$$

$$6 - 2 \cdot (-1\ 1/2) \stackrel{?}{=} 9$$
$$6 + 3 \stackrel{?}{=} 9 \checkmark$$

5. Solve. Check your solutions (as always!).

a. $1 - 5x = 2$

b. $12 - 3y = -6$

c. $10 = 8 - 4y$

d. $7 = 5 - 3t$

6. Choose from the expressions at the right to build an equation that has the root $x = 2$.

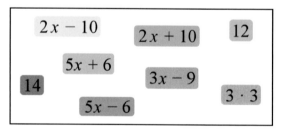

$2x - 10 \quad 2x + 10 \quad 12$
$5x + 6$
$14 \quad 3x - 9$
$5x - 6 \quad 3 \cdot 3$

7. Choose from the expressions at the right to build an equation that has the root $x = 5$.

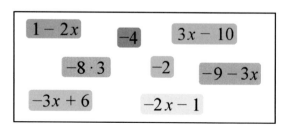

$1 - 2x \quad -4 \quad 3x - 10$
$-8 \cdot 3 \quad -2$
$-3x + 6 \quad -9 - 3x$
$-2x - 1$

70

Example 5. Solve as a decimal.

$$\frac{3x}{7} = 0.9 \quad \Big| \cdot 7$$

$$\frac{\cancel{7} \cdot 3x}{\cancel{7}} = 6.3$$

$$3x = 6.3 \quad \Big| \div 3$$

$$x = 2.1$$

Check:
$$\frac{3 \cdot 2.1}{7} \stackrel{?}{=} 0.9$$

$$\frac{6.3}{7} \stackrel{?}{=} 0.9$$

$$0.9 = 0.9 \checkmark$$

In this example, you *could* first divide by 3 and then multiply by 7. The solution wouldn't be any more difficult that way.

8. Solve.

a. $\dfrac{2x}{5} = 8$	b. $\dfrac{3x}{8} = -9$
c. $15 = \dfrac{-3x}{10}$	d. $2 - 4 = \dfrac{s+4}{5}$
e. $\dfrac{x}{2} - (-16) = -5 \cdot 3$	f. $2 - 4p = -0.5$

Two-Step Equations: Practice

Example 1. The number line diagram illustrates the equation $-29 + x + 7 = -4$:

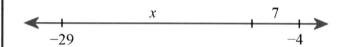

Think of starting at -29, jumping x steps, jumping another 7 steps, and arriving at -4.

You can find the value of the unknown x using logical thinking or by writing an equation and solving it. Here is the solution using an equation:

$$-29 + x + 7 = -4 \quad \text{(add } -29 + 7 \text{ on the left side)}$$
$$-22 + x = -4$$
$$+22 \quad +22$$
$$x = 18$$

1. Write an equation to match the number line model and solve for the unknown.

a.

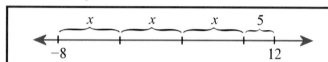

b.

2. Solve. Compare equation (a) to equation (b). They are similar, yet different!

a. $2 - 5y = -11$

b. $5y - 2 = -11$

3. Solve. Compare equation (a) to equation (b). They are similar, yet different!

a. $\dfrac{2x}{7} = -5$	b. $\dfrac{x+2}{7} = -5$

4. Solve. Check your solutions (as always!).

a. $20 - 3y = 65$	b. $6z + 5 = -2.2$
c. $\dfrac{t+6}{-2} = -19$	d. $\dfrac{y}{6} - 3 = -0.7$

Example 2. The perimeter of an isosceles triangle is 26 inches and its base measures 5 inches. How long are the two sides that are equal (congruent)?

(To help yourself, label the image with the data from the problem.)

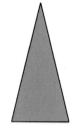

Read the two solutions below. Notice how neatly they tie in with each other!

Solution 1: an equation	Solution 2: Logical thinking/mental math
Let x be the unknown side length. We get: $$\text{perimeter} = x + x + 5$$ $$26 = 2x + 5$$ Next we solve the equation: $$2x + 5 = 26 \quad \vert -5$$ $$2x = 21 \quad \vert \div 2$$ $$x = 10.5$$ The two other sides measure 10.5 inches each.	The perimeter is 26 inches. This means that the two unknown sides and the 5-inch side add up to 26 inches. Therefore, if we subtract the base side, the two unknown sides must add up to 21 inches. So one side is half of that, or 10.5 inches.

5. Solve each problem below in two ways: write an equation, and use logical reasoning/mental math.

a. A quadrilateral has three congruent sides. The fourth side measures 1.4 m. If the perimeter of the quadrilateral is 7.1 meters, what is the length of each congruent side?	
Equation:	Mental math/logical thinking:

b. You bought six identical baskets from an artisan. She gave you a $12 discount on your order, and your total bill was only $46.80. What is the normal price of one basket?

Equation:

Logical thinking:

c. Two-fifths of a number is 466. What is the number?

Equation:

Logical thinking:

Growing Patterns 2

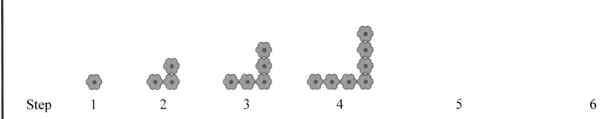

| Step | 1 | 2 | 3 | 4 | 5 | 6 |

How do you think this pattern is growing?

How many flowers will there be in step 39?

This pattern adds 2 flowers in each step, except in step 1. This means that by step 39, we have added 2 flowers 38 times. Therefore, there are $1 + 2 \cdot 38 = 77$ flowers in step 39.

Write a formula for the number of flowers in step n.

There are several ways to do this. The three ways explained below are not the only ones!

1. Let's view the pattern as adding 2 flowers in each step after the first one. By step n, the pattern has added one less than n times 2 flowers, because we need to exclude that first step. This means that $(n - 1)$ times 2, or $(n - 1) \cdot 2$, flowers added to the one flower that we started with.

   This gives us the expression $1 + (n - 1) \cdot 2$. Since we customarily put the variable first and the constant last, we can rewrite that expression as $1 + 2(n - 1)$ and then as $2(n - 1) + 1$.

2. Another way to think about this pattern is as two legs. One leg includes the flower in the corner, so it has the same number of flowers as the step number. The other leg doesn't have the corner flower, so it has one flower less than the step numbers. In other words, in step 3, we have $3 + 2$ flowers. In step 4, we have $4 + 3$ flowers. In step 5, we have $5 + 4$ flowers.

   This gives us a formula for the number of flowers in step n: there are $n + (n - 1)$ flowers in step n.

3. Yet another way is that, in each step, there are twice as many flowers as the step number, minus one for the flower that is shared. For example, in step 4, we have twice 4 minus 1, which is seven flowers.

   This also gives us a formula: there are $2n - 1$ flowers in step n.

All of the formulas are equivalent (just as we would expect!) and simply represent different ways of thinking about the number of flowers in each step. On the right, you can see how the first two formulas can be simplified to the third one.

$n + (n - 1)$ $2(n - 1) + 1$
$= n + n - 1$ $= 2n - 2 + 1$
$= 2n - 1$ $= 2n - 1$

In which step are there 583 flowers?

We can use our formula to write an equation to answer this question. In the question the step number n is unknown, but the total number of flowers in that step is 583. Since we know from our formula that there are $2n - 1$ flowers in step n, we get

$$\begin{aligned} 2n - 1 &= 583 \quad &+1 \\ 2n &= 584 \quad &\div 2 \\ n &= 292 \end{aligned}$$

1   2   3   4   5

1. **a.** How is this pattern growing?

   b. How many triangles will there be in step 39?

   c. Write a formula for the number of triangles in step n.
   Check your answer with your teacher before going on to part (d).

   d. In which step will there be 311 triangles?
   Write an equation and solve it.
   Notice, this question is different from the one in part (c).

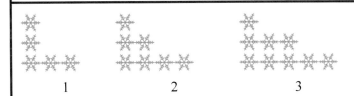

    1   2   3   4   5

2. **a.** How do you think this pattern is growing?

   b. How many snowflakes will there be in step 39?

   c. Write a formula for the number of snowflakes in step n.
   Check your answer with your teacher before going on to part (d).

   d. In which step will there be 301 snowflakes?
   Write an equation and solve it.

> Instead of showing the steps of the pattern horizontally, like this...
>
> 1 2 3 4
>
> ...we can also show them like this:
>
> 1
> 2
> 3
> 4
> ...
>
> Now, each row of flowers is one step of the pattern.

3. A section of a flower garden has rows of flowers. The first row has four flowers, and each row after that has one more flower than the previous row.

   1
   2
   3
   ...

Row	Flowers
1	3 + 1
2	3 + 2
3	3 + 3
4	3 + 4
n	

   a. Write a formula that tells the gardener the number of flowers in row n.

   b. How many flowers are in the 28th row?

   c. In which row will there be 97 flowers? Write an equation and solve it.

4. This pattern is similar to the previous one. This time each row has *two* more flowers than the previous row. Notice that the number of flowers in each row gives us the list of all the odd numbers.

   0
   1
   2
   3
   ...

   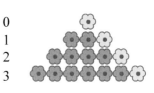

   Here's one way to look at this pattern: Label the first row as "row 0." Then the number of flowers in any row is twice the row number plus 1.

   a. Write a formula that tells the gardener the number of flowers in row n.

   b. In which row will there be 97 flowers? Write an equation and solve it.

5. Each floor of a multi-story building is 9 feet high.

   a. Write an expression for the total height of the building, if it is n stories high.

   b. Write an expression for the total height of the building if it is n stories high, and the bottom of the first floor is elevated 2 ft above the ground.

   c. How many stories does the building have if the total height of the building is 164 ft? Solve this problem in two ways: using logical thinking and using an equation.

6. Jeremy earns $400 a week. He also earns $15 for every hour he works overtime.

   a. Write an expression for Jeremy's *total* earnings if he works n hours overtime. You can use the table to help you.

   b. How many hours should Jeremy work overtime in order to earn $970 in a week?

Overtime hours	Total earnings
0	
1	
2	
17	
n	

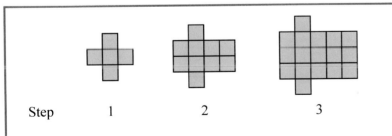

Step   1   2   3   4   5

What is the pattern of growth here?
How many squares will there be in step 59?

Puzzle Corner

Review 2

1. Solve. Check your solutions.

a.	$x + 7 = -6$	b.	$-x = 5 - 9$
c.	$2 - x = -8$	d.	$2 - 6 = -z + 5$
e.	$\dfrac{x}{11} = -12$	f.	$\dfrac{q}{-3} = -40$
g.	$100 = \dfrac{c}{-10}$	h.	$\dfrac{a}{5} = -10 + (-11)$

Write an equation for the problem. Then solve it.

2. Alex bought three identical solar panels and paid a total of $837. How much did one cost?

   Equation:

Write an equation for the problem. Then solve it.

3. Andrew pays 1/7 of his salary in taxes. If he paid $187 in taxes, how much was his salary?

   Equation:

4. Use the formula $d = vt$ to solve the problem.

If you can bicycle at a speed of 20 km/h, how long will it take you to bicycle from the shopping center to a dentist's office, a distance of 1.2 km?	d = v t ↓ ↓ ↓

5. Taking a bus, Emily can get to the community center that is 1.5 km from her home in 3 minutes. What is the average speed of the bus, in kilometers per hour?

6. Ed skates on his skateboard to school, which is 2 miles away. He travels half of the distance at a speed of 12 mph and the rest at a speed of 15 mph. How long does it take him to get to school?

Get Ready for Algebra: Grade 7 Answer Key

Order of Operations, p. 17

1. a. 64   b. 100,000   c. 0.01
   d. 0.008   e. 1   f. 1,000,000

2. a. $0^5 = 0$
   b. $0.9^2 = 0.81$
   c. $5^3 + 2^5 = 125 + 32 = 157$
   d. $6 \cdot 10^6 - 9 \cdot 10^4 = 6{,}000{,}000 - 90{,}000 = 5{,}910{,}000$

3. a. 8 cm^3 b. 121 ft^2 c. 1.44 km^2

4. a. (i) 8 cm^3 b. (ii) (8 cm)^3

5.

a. 95	c. 100	e. 2,000
b. 95	d. 400	f. 210
g. 0	h. 0.343	i. 245

6. a. 35   b. 9   c. 49   d. 4.2

7. a. 30   b. 13   c. 9

8. a. $\frac{2}{5} \cdot 4 = 1\ 3/5$

   b. $\frac{16}{2+6} \cdot 2 = 4$

   c. $4 + \frac{1}{3} + 2 = 6\ 1/3$

   d. $\frac{2}{3 \cdot 4} = 1/6$

   e. $\frac{5}{9} \cdot 3 + 1 = 2\ 2/3$

   f. $\frac{1+3}{4+2} = 2/3$

9. a. 54   b. 8/9   c. 10 4/9

10. a. 28   b. 28   c. 2 1/4
    d. 1 1/5   e. 3 1/3   f. 4 3/8

11. a. cost = $4n + 6m$
    b. value in cents = $1x + 10y$
    c. Mike's cookies = $10 - t$
    d. earnings = $\$11n$
    e. girls = $\frac{2}{3} \cdot 81 = 54$
    f. girls = $\frac{2}{3} n$
    g. price = $\frac{9}{10} \cdot \$60 = \$54$
    h. price = $\frac{9}{10} x$ or price = $0.9x$
    i. $A = 3b \div 2$
    j. $V = c^3$

Expressions and Equations, p. 21

1. a. $2x - (40 + x)$ b. $(3x)^3$ c. $s - 6$
   d. $5b^5$ e. $7(x - y)$ f. $t^2 - s^2$
   g. $2^3 - x$ h. $5 \div y^2$ i. $x^5 - 2$
   j. $x^3 \cdot y^2$ k. $(2x + 1)^4$ l. $\frac{x - y}{x^2 + 1}$

2.

Equation	Solution
a. $78 - x = 8$	$x = 70$
b. $x - 2/3 = 1/4$	$x = 11/12$
c. $x \div 7 = 3/21$	$x = 1$

3. a. 2 and 7
   b. 5/7.

4. 3/2 makes the equation true.

5. a. $5 + n$ or $n + 5$
   b. $A = B - 8$

6. a. There are $2x$ red blocks.
   b. There are $3x$ green blocks.

7. a. and b.
   Since Timothy gets to keep 4/5 of his salary, the expression is $4/5s$. Another correct expression is $s - (1/5)s$. Using decimals, the correct expressions are $0.8s$ and $s - 0.2s$.

8. a. $\frac{3p}{4} = \$57$

   b. $3 \cdot 25 + 5p = 98$ or $5p + 75 = 98$

   c. $3(x - \$1) = \$5.40$

Puzzle corner:
$4 + c + 1 = 2 + 3 + c$ and $3z - 1 = z - 1 + z + z$.

Properties of the Four Operations, p. 24

1. a. Yes, they are the same.
   b. No, they are not the same.
   c. No, they are not the same.
   d. Yes, they are the same.

2. No, subtraction is not commutative. The counterexamples will vary, so check the student's counterexample. For example, if $a = 3$ and $b = 5$, then $3 - 5 = -2$ is not the same as $5 - 3 = 2$.

3. No, division is not commutative. The counterexamples will vary, so check the student's work. For example, if $a = 10$ and $b = 4$, then $10 \div 4 = 2.5$ is not the same as $4 \div 10 = 0.4$.

4. a. Yes   b. Yes   c. No.

5. a. The associative property of addition.
   b. The associative property of multiplication.

6. b. Yes
   c. No. For example, when n is 5 and m is 1, we get:

   $(n - 2) \cdot m \stackrel{?}{=} m(2 - n)$
   $(5 - 2) \cdot 1 \stackrel{?}{=} 5(2 - 1)$
   $3 \neq -3$

   d. No, the expressions are not equal.
   For example, when a is 5 and b is 1, we get:

   $a + 2b \stackrel{?}{=} b + 2a$
   $5 + 2 \cdot 1 \stackrel{?}{=} 1 + 2 \cdot 5$
   $7 \neq 11$

7. No. For example, let $a = 15$, $b = 2$, and $c = 7$. Then, $(a - b) - c = (15 - 2) - 7 = 6$ and $a - (b - c) = 15 - (2 - 7) = 15 - (-5) = 20$. Since the expressions are not equal, subtraction is not associative.

8. No. For example, let $a = 10$, $b = 5$, and $c = 2$. Then, $(a \div b) \div c = (10 \div 5) \div 2 = 1$ and $a \div (b \div c) = 10 \div (5 \div 2) = 4$. Since the expressions are not equal, division is not associative.

9. a. $x + x + x + x + x = 5x$
   b. $a + a + b + b + a = 3a + 2b$
   c. $s + s + 15 + s + s = 4s + 15$
   d. $v + 11 + v + 16 + v + t = t + 3v + 27$

10. a. $3v + 13$
    b. $4e + 37$
    c. $3v$
    d. $13a$
    e. $15a + 8b + 8$
    f. $5s + 10t + 2$

11. a. $7a^2$
    b. $16s^2$
    c. a^2d^4
    d. $2(xy)^3$
    e. $2a + 2d$
    f. $4z + 8$
    g. $8t^2y^3$
    h. $27b^4$
    i. $6s + t + 2$

12. a. No. For example, 5/3 is not equal to 3/5.
    b. No. For example, $3 + \dfrac{4}{2}$ is not equal to $4 + \dfrac{3}{2}$.
    c. Yes.
    d. Yes.

13.

Operation	Commu-tative?	Associ-ative?	notes/examples
addition	yes	yes	
subtraction	no	no	$5 - 3 \neq 3 - 5$ and $(9 - 5) - 3 \neq 9 - (5 - 3)$
multiplication	yes	yes	
division	no	no	$5 \div 3 \neq 3 \div 5$ and $(8 \div 4) \div 2 \neq 8 \div (4 \div 2)$

Simplifying Expressions, p. 28

1. a. $3p + 8$
   b. $8p^4$
   c. $6p$
   d. $8p^2$
   e. $10x^3$
   f. $12y^4$

2.
a. Area = x^2 Perimeter = $4x$	b. Area = $4x^2$ Perimeter = $8x$
c. Area = $8x^2$ Perimeter = $12x$	d. Area = $9x^2$ Perimeter = $12x$

3. a.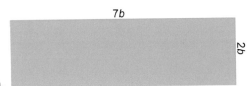
   b. Area = $14b^2$
   c. Perimeter = $18b$

4. a. The answers will vary. Check the student's drawing and answers. The table gives some possibilities for the lengths of the sides and the resulting areas.

side A	side B	area
$11s$	$1s$	$11s^2$
$10s$	$2s$	$20s^2$
$9s$	$3s$	$27s^2$
$8s$	$4s$	$32s^2$
$7s$	$5s$	$35s^2$
$6s$	$6s$	$36s^2$
$6.5s$	$5.5s$	$35.75s^2$
$8.2s$	$3.8s$	$31.16s^2$
$(14/3)s$	$(22/3)s$	$(308/9)s^2$

   b. Answers will vary. Check the student's drawing. For example, if the student's rectangle in part (a) had sides $8s$ and $4s$, then the sides are 24 cm and 12 cm, and the area is 288 cm^2.

5. a. The expression $2a \cdot 2b$ is for area.
      The expression $4a + 4b$ is for perimeter.
   b. Answers will vary. Check the student's drawing and answers. For example:

6. a.
Value of p	$3p$	$p + 3$
0	0	3
0.5	1.5	3.5
1	3	4
1.5	4.5	4.5
2	6	5
2.5	7.5	5.5
3	9	6
3.5	10.5	6.5
4	12	7

   b. No, you can't. It varies. For some values of p, $3p > p + 3$. For others, it is just the opposite.

7.
Expression	The terms in it	Coefficient(s)	Constants
$(5/6)s$	$(5/6)s$	5/6	none
w^3	w^3	1	none
$0.6x + 2.4y$	$0.6x$ and $2.4y$	0.6 and 2.4	none
$x + 3y + 7$	x and $3y$ and 7	1 and 3	7
$p \cdot 101$	$101p$	101	none
$x^5y^2 + 8$	x^5y^2 and 8	1	8

8. a. $12p$
   b. $12p^2$
   c. a^2

9. a. Area = $8b \cdot 2b + 6b \cdot 3b = 16b^2 + 18b^2 = 34b^2$
   b. Area = $4x \cdot 6x + 2x \cdot 2x = 24x^2 + 4x^2 = 28x^2$

10. a. Expression: $8 + 2w$
    b. Equation: $8 + 2w = 22$. Solution: $w = 7$.
       The width is 7 meters.

11. a. Expression: $8w$
    b. Equation: $8w = 216$. Solution: $w = 27$.
       In 27 weeks they will have borrowed 216 books.

12. a. Expression: $6y + 5$
    b. Equation: $6y + 5 = 155$. Solution: $y = 25$.
       She bought 25 containers of mints.

Simplifying Expressions, cont.

Puzzle corner.
a. $10n + 25m$
b. There are ten possibilities (see the table).

Dimes	Value of dimes (cents)	Quarters	Value of quarters (cents)	Total value (cents)
2	20	19	475	495
7	70	17	425	495
12	120	15	375	495
17	170	13	325	495
22	220	11	275	495
27	270	9	225	495
32	320	7	175	495
37	370	5	125	495
42	420	3	75	495
47	470	1	25	495

Growing Patterns 1, p. 32

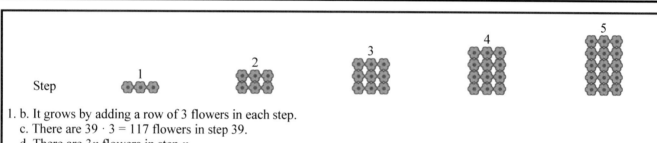

1. b. It grows by adding a row of 3 flowers in each step.
    c. There are $39 \cdot 3 = 117$ flowers in step 39.
    d. There are $3n$ flowers in step n.

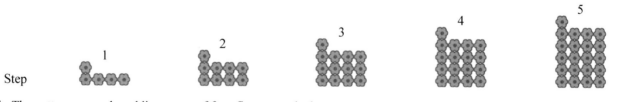

2. b. The pattern grows by adding a row of four flowers to the bottom.
    c. There are $39 \cdot 4 + 1 = 157$ flowers in step 39.
    d. There are $4n + 1$ flowers in step n. Another formula: There are $4(n + 1) - 3$ flowers in step n.

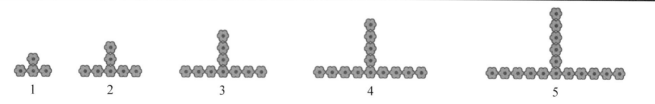

3. b. In each step, one flower is added on the left, one flower on the right, and one flower on top.
    c. There are $38 \cdot 3 + 4 = 118$ flowers in step 39.
    d. There are $3(n - 1) + 4$ flowers in step n. Another formula: There are $3n + 1$ flowers in step n.

Growing Patterns 1, cont.

Step    1   2   3   4   5

4. b. Each step adds a row of three flowers on the bottom.
   c. There are $39 \cdot 3 + 2 = 119$ flowers in step 39.
   d. There are $3n + 2$ flowers in step n. Another formula: There are $5 + 3(n-1)$ flowers in step n.

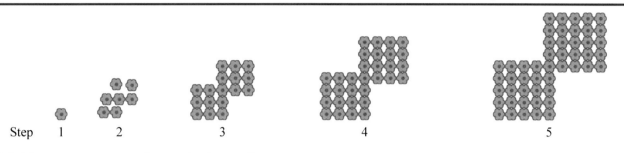

5. b. There are two squares that overlap by one flower.
   c. There are $39^2 + 39^2 - 1 = 3{,}041$ flowers in step 39.
   d. There are $n^2 + n^2 - 1 = 2n^2 - 1$ flowers in step n.

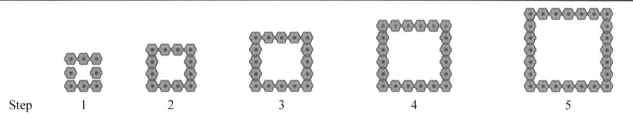

6. b. The length of each side is expanded by one flower in each step.
   c. There are $4 \cdot 39 + 4 = 160$ flowers in step 39.
   d. There are $4n + 4$ flowers in step n.

The Distributive Property, p. 35

1. a. $3(2a + 8) = 6a + 24$
   b. $4(s + 11) = 4s + 44$
   c. $2(3z + 7) = 6z + 14$

2. a. $3(b + 8) = 3b + 24$

   b. $4(2w + 1) = 8w + 4$

   c. $2(3x + 5) = 6x + 10$

   | x | x | x | 5 | x | x | x | 5 |

3. a. $2x + 18$
   b. $28y + 35$
   c. $90x + 80$
   d. $8x + 8y$
   e. $4s + st$
   f. $uv + uw$

The Distributive Property, cont.

4. The perimeter equals $6(2x + 4) = 12x + 24$

5. Each side is $¼ \cdot (24y + 40) = 6y + 10$.

6. a. $20n$
   b. $20n + 11$
   c. $3(20n + 11) = 60n + 33$
   d. You bought 18 jars, or 6 jars in each of the three orders. You can work backwards: if three orders total $393, then one order is $131. Of that, $11 is the shipping fee, so the jars of coconut oil cost $120. This means you purchased 6 jars in one order. Therefore, you bought 18 jars in total (in three orders). If you know how to solve equations, you can solve the equation $60n + 33 = 393$, and get $n = 6$.

7. a. $11x - 77$
   b. $30x + 30y + 150$
   c. $10r + 20s + 1$
   d. $15x - 10y - 30$
   e. $1.5s + st - sx$
   f. $1.5v + w - 3.5$

8. a. 792   b. 588   c. 2,995

9.

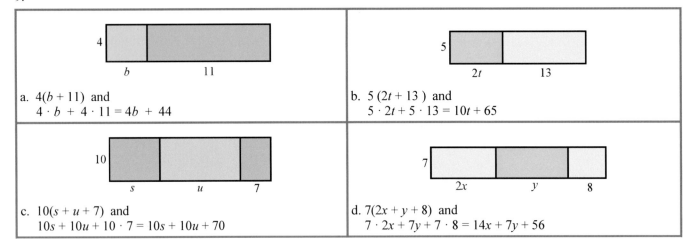

a. $4(b + 11)$ and
$4 \cdot b + 4 \cdot 11 = 4b + 44$

b. $5(2t + 13)$ and
$5 \cdot 2t + 5 \cdot 13 = 10t + 65$

c. $10(s + u + 7)$ and
$10s + 10u + 10 \cdot 7 = 10s + 10u + 70$

d. $7(2x + y + 8)$ and
$7 \cdot 2x + 7y + 7 \cdot 8 = 14x + 7y + 56$

10.

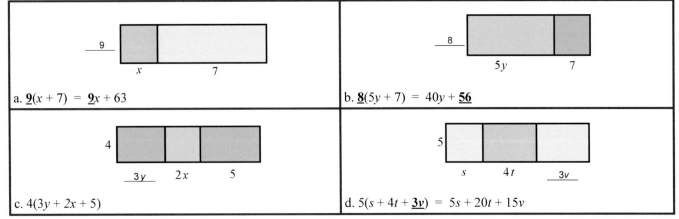

a. $\underline{9}(x + 7) = \underline{9}x + 63$

b. $\underline{8}(5y + 7) = 40y + \underline{56}$

c. $4(3y + 2x + 5)$

d. $5(s + 4t + \underline{3v}) = 5s + 20t + 15v$

11.

a. $\underline{10}(20x + 3) = 200x + 30$	b. $4(6s - \underline{x}) = 24s - 4x$
c. $2(\underline{3.5x} + 1.5y + 0.9) = 7x + 3y + 1.8$	d. $4(\underline{0.1x} - \underline{0.3y} + \underline{0.4}) = 0.4x - 1.2y + 1.6$

The Distributive Property, cont.

12.

a. $2x + 6 = \underline{2}(x + 3)$	b. $4y + 16 = 4(\underline{y + 4})$
c. $21t + 7 = \underline{7}(3t + \underline{1})$	d. $16d + 24 = \underline{8}(2d + \underline{3})$
e. $15x - 35 = \underline{5(3x - 7)}$	f. $7a - 49 = \underline{7(a - 7)}$

13. a. 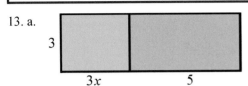   b.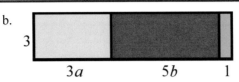

14.

a. $64x + 40 = 8(8x + 5)$	b. $54x + 18 = 18(3x + 1)$
c. $100y - 20 = 20(5y - 1)$	d. $90t + 33s + 30 = 3(30t + 11s + 10)$
e. $36x - 12y + 24 = 12(3x - y + 2)$	f. $2x + 8z - 40 = 2(x + 4z - 20)$

15. a. $\dfrac{300 + 2}{3} = \dfrac{300}{3} + \dfrac{2}{3} = 100\dfrac{2}{3}$

b. $\dfrac{13 + 700}{7} = \dfrac{13}{7} + \dfrac{700}{7} = 101\dfrac{6}{7}$

c. $\dfrac{5{,}031}{5} = \dfrac{5{,}000}{5} + \dfrac{31}{5} = 1{,}006\dfrac{1}{5}$

d. $\dfrac{5x - 3}{6} = \dfrac{5x}{6} - \dfrac{3}{6} = \dfrac{5}{6}x - \dfrac{1}{2}$

e. $\dfrac{x + 7}{7} = \dfrac{x}{7} + \dfrac{7}{7} = \dfrac{x}{7} + 1$

f. $\dfrac{4x + 2}{4} = \dfrac{4x}{4} + \dfrac{2}{4} = x + \dfrac{1}{2}$

16. a. The house can be 33 ft long. You can see from the image on the right that the garage is 25 by 15 ft, which is 375 sq. ft. Subtract 1200 sq. ft. − 375 sq. ft. = 825 sq. ft. to find the area of the actual house. Lastly divide 825 sq. ft. ÷ 25 ft = 33 ft.
b. $25(x + 15) = 1200$ or $(x + 15)25 = 1200$ or similar.

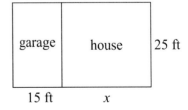

Puzzle corner:

a. $5(y - 9) = 5 \cdot y - 5 \cdot 9$

b.

Review 1, p. 40

1. a. 2,700   b. 9   c. 7 1/2

2. The associative property of multiplication.

3. a. 50   b. 8/67

4. a. $l_1 = l_2 - 1.5$

   b. $\dfrac{5w}{6} = 23$

5. No. For example, $5 - 3$ is not the same as $3 - 5$.

6. a. $p - p/5$ or $p - (1/5)p$ or $(1 - 1/5)p$ or $(4/5)p$ or $0.8p$
   b. $0.18y$
   c. $3n + 8m$

7. a. $3x + 2$
   b. $2x^4$
   c. $20v$
   d. $96v^2$
   e. $36z^3$
   f. $2f + 6g$

8. a.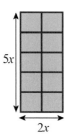
   b. The area is $10x^2$.
   c. The perimeter is $14x$.

9. a. $12v - 108$
   b. $3a + 3b + 6$
   c. $1.5t - 3x$

10. [rectangle divided into two parts, height 11, widths x and 7]

11.

Expression	The terms in it	Coefficient(s)	Constants
a^8	a^8	1	none
$2x + 9y$	$2x$ and $9y$	2 and 9	none

12. The side is $6s + 9$.

13. a. $48x + 12 = 12(4x + 1)$
    b. $40x - 25 = 5(8x - 5)$
    c. $6y - 2z = 2(3y - z)$
    d. $56t - 16s + 8 = 8(7t - 2s + 1)$

Solving Equations, p. 42

1.

a.
Equation	Operation to do to both sides
$x + 1 = -4$	-1
$x + 1 + (-1) = -4 + (-1)$	
$x = -5$	

b.
Equation	Operation to do to both sides
$x - 1 = -3$	$+1$
$x + (-1) + 1 = -3 + 1$	
$x = -2$	

c.
Equation	Operation to do to both sides
$x - 2 = 6$	$+2$
$x + (-2) + 2 = 6 + 2$	
$x = 8$	

d.
Equation	Operation to do to both sides
$x + 5 = 2$	-5
$x + 5 + (-5) = 2 + (-5)$	
$x = -3$	

2. a. $x = 10$ b. $x = -1$ c. $x = -7$ d. $x = -9$

3.

a.
Equation	Operation to do to both sides
$4x = -12$	$\div 4$
$4x \div 4 = -12 \div 4$	
$x = -3$	

b.
Equation	Operation to do to both sides
$(1/3)x = -5$	$\cdot 3$
$(1/3)x \cdot 3 = -5 \cdot 3$	
$x = -15$	

4. a. $p_l = p_e - 8$

b. $\dfrac{4p}{5} = \$16$

5. $x = -3$

6. Students can solve these equations using guess and check.

a. $x - 7 = 5$
   $x = 12$

b. $5 - 8 = x + 1$
   $x = -4$

c. $\dfrac{x - 1}{2} = 4$
   $x = 9$

d. $x^3 = 8$
   $x = 2$

e. $-3 = \dfrac{15}{y}$
   $y = -5$

f. $5(x + 1) = 10$
   $x = 1$

7.

a.
Equation	Operation to do to both sides
$3x + 1 = -5$	
$3x + 1 - 1 = -5 - 1$	-1
$3x = -6$	
$3x \div 3 = -6 \div 3$	$\div 3$
$x = -2$	

b.
Equation	Operation to do to both sides
$2x - 3 = 4$	
$2x - 3 + 3 = 4 + 3$	$+3$
$2x = 7$	
$2x \div 2 = 7 \div 2$	$\div 2$
$x = 3\ 1/2$	

c.
Equation	Operation to do to both sides
$4x + 1 = 13$	
$4x + 1 - 1 = 13 - 1$	-1
$4x = 12$	
$4x \div 4 = 12 \div 4$	$\div 4$
$x = 3$	

Addition and Subtraction Equations, p. 47

1.

a.	$x + 5 = 9$ $-5 -5$ $x = 4$	b.	$x + 5 = -9$ $-5 -5$ $x = -14$
c.	$x - 2 = 3$ $+2 +2$ $x = 5$	d.	$w - 2 = -3$ $+2 +2$ $w = -1$
e.	$z + 5 = 0$ $-5 -5$ $z = -5$	f.	$y - 8 = -7$ $+8 +8$ $y = 1$

2.

a.	$x - 7 = 2 + 8$ $x - 7 = 10$ $+7 +7$ $x = 17$	b.	$x - 10 = -9 + 5$ $x - 10 = -4$ $+10 +10$ $x = 6$
c.	$s + 5 = 3 + (-9)$ $s + 5 = -6$ $-5 -5$ $s = -11$	d.	$t + 6 = -3 - 5$ $t + 6 = -8$ $-6 -6$ $t = -14$

Addition and Subtraction Equations, cont.

3.

a.	$-8 = s + 6$ $s + 6 = -8$ $\underline{-6 \quad -6}$ $s = -14$ Check: $-8 \stackrel{?}{=} -14 + 6$ $-8 = -8$ ✓	b.	$-2 = x - 7$ $x - 7 = -2$ $\underline{+7 \quad +7}$ $x = 5$ Check: $-2 \stackrel{?}{=} 5 - 7$ $-2 = -2$ ✓
c.	$4 = s + (-5)$ $s + (-5) = 4$ $\underline{+5 \quad +5}$ $s = 9$ Check: $4 \stackrel{?}{=} 9 + (-5)$ $4 = 4$ ✓	d.	$2 - 8 = y + 6$ $y + 6 = -6$ $\underline{-6 \quad -6}$ $y = -12$ Check: $2 - 8 \stackrel{?}{=} -12 + 6$ $-6 = -6$ ✓
e.	$5 + x = -9$ $x + 5 = -9$ $\underline{-5 \quad -5}$ $x = -14$ Check: $5 + (-14) \stackrel{?}{=} -9$ $-9 = -9$ ✓	f.	$-6 - 5 = 1 + z$ $z + 1 = -11$ $\underline{-1 \quad -1}$ $z = -12$ Check: $-6 - 5 \stackrel{?}{=} 1 + (-12)$ $-11 = -11$ ✓
g.	$y - (-7) = 1 - (-5)$ $y + 7 = 6$ $\underline{-7 \quad -7}$ $y = -1$ Check: $-1 - (-7) \stackrel{?}{=} 1 - (-5)$ $6 = 6$ ✓	h.	$6 + (-2) = x - 2$ $x - 2 = 4$ $\underline{+2 \quad +2}$ $x = 6$ Check: $6 + (-2) \stackrel{?}{=} 6 - 2$ $4 = 4$ ✓
i.	$3 - (-9) = x + 5$ $x + 5 = 12$ $\underline{-5 \quad -5}$ $x = 7$ Check: $3 - (-9) \stackrel{?}{=} 7 + 5$ $12 = 12$ ✓	j.	$2 - 8 = 2 + w$ $w + 2 = -6$ $\underline{-2 \quad -2}$ $w = -8$ Check: $2 - 8 \stackrel{?}{=} 2 + (-8)$ $-6 = -6$ ✓

Addition and Subtraction Equations, cont.

4.

a.	$-x = 6$ $x = -6$	b.	$-x = 5 - 9$ $-x = -4$ $x = 4$
c.	$4 + 3 = -y$ $-y = 7$ $y = -7$	d.	$-2 - 6 = -z$ $-z = -8$ $z = 8$

5. a. $4(s - 1/2) = 12$ and $4(s - 0.5) = 12$
   b. Since $4 \times 3 = 12$, $s - ½ = 3$. So the sides were $3 + ½ = 3½$ meters long before they were shortened.

   c. $\quad\quad 4(s - 1/2) = 12$ OR c. $\quad\quad 4(s - 1/2) = 12$
   $\quad\quad\quad\quad s - 1/2 = 12/4$ $\quad\quad\quad\quad\quad\quad\quad\quad\quad\quad 4s - 2 = 12$
   $\quad\quad\quad\quad s - 1/2 = 3$ $\quad\quad\quad\quad\quad\quad\quad\quad\quad\quad 4s - 2 + 2 = 12 + 2$
   $\quad\quad s - 1/2 + 1/2 = 3 + 1/2$ $\quad\quad\quad\quad\quad\quad\quad\quad\quad\quad 4s = 14$
   $\quad\quad\quad\quad\quad\quad s = 3\ 1/2$ $\quad\quad\quad\quad\quad\quad\quad\quad\quad\quad s = 3\ 1/2$

   This solution matches this thought process:

   Right now, the perimeter is 12 m, so each side is $12 \div 4 = 3$ m. So, before the sides were reduced by $1/2$ m, they were $3\ 1/2$ m long.

   This solution matches this thought process:

   If we add the $1/2$ meter back in to each side, the perimeter increases by a total of 2 m and becomes 14 meters. So the length of the original side is that 14 m perimeter divided by 4, which is $3½$ m.

6.

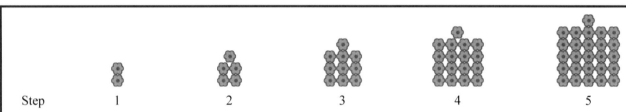

Step 1 2 3 4 5

a. In each step, the sides of the square increase by 1 flower. The one flower on the top does not change.
b. There will be 1,522 flowers in step 39.
c. There will be $n^2 + 1$ flowers in step n.

Addition and Subtraction Equations, cont.

7.

a. $2 + (-x) = 6$ $\quad\mid -2$ $-x = 4$ $x = -4$ Check: $2 + (-(-4)) \stackrel{?}{=} 6$ $2 + 4 = 6$ ✓	b. $8 + (-x) = 7$ $\quad\mid -8$ $-x = -1$ $x = 1$ Check: $8 + (-1) \stackrel{?}{=} 7$ $7 = 7$ ✓	
c. $-5 + (-x) = 5$ $\quad\mid +5$ $-x = 10$ $x = -10$ Check: $-5 + (-(-10)) \stackrel{?}{=} 5$ $-5 + 10 = 5$ ✓	d. $2 + (-x) = -6$ $\quad\mid -2$ $-x = -8$ $x = 8$ Check: $2 + (-8) \stackrel{?}{=} -6$ $-6 = -6$ ✓	
e. $1 = -5 + (-x)$ $\quad\mid +5$ $-x = 6$ $x = -6$ Check: $1 \stackrel{?}{=} -5 + (-(-6))$ $1 = -5 + 6$ ✓	f. $2 + (-9) = 8 + (-z)$ $\quad\mid -8$ $-z = -15$ $z = 15$ Check: $2 + (-9) \stackrel{?}{=} 8 + (-15)$ $-7 = -7$ ✓	
g. $-8 + r = -5 + (-7)$ $\quad\mid +8$ $r = -4$ Check: $-8 + (-4) \stackrel{?}{=} -5 + (-7)$ $-12 = 12$ ✓	h. $2 - (-5) = 2 + 5 + t$ $7 = 7 + t$ $\quad\mid -7$ $0 = t$ Check: $2 - (-5) \stackrel{?}{=} 2 + 5 + 0$ $7 = 7$ ✓	

Multiplication and Division Equations, p. 53

1.

a. $\dfrac{8x}{8} = x$	b. $\dfrac{8x}{2} = 4x$	c. $\dfrac{2x}{8} = \dfrac{x}{4}$
d. $\dfrac{-6x}{-6} = x$	e. $\dfrac{-6x}{6} = -x$	f. $\dfrac{6x}{-6} = -x$
g. $\dfrac{6w}{2} = 3w$	h. $\dfrac{6w}{w} = 6$	i. $\dfrac{6w}{-2} = -3w$

Multiplication and Division Equations, cont.

2.

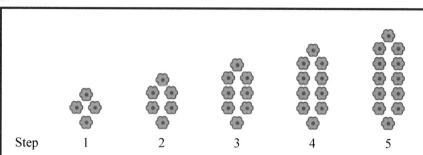

Step   1   2   3   4   5

a. The sides of the shape get one flower taller in each step.
b. There will be 80 flowers in step 39.
c. There will be $2n + 2$ flowers in step n.

3.

a. $\quad 5x = -45 \quad \mid \div 5$ $\quad 5x/5 = -45/5$ $\quad \mathbf{x = -9}$ Check: $5 \cdot (-9) \stackrel{?}{=} -45$ $\qquad -45 = -45$ ✓	b. $\quad -3y = -21 \quad \mid \div (-3)$ $\quad -3y/(-3) = -21/(-3)$ $\quad \mathbf{y = 7}$ Check: $-3 \cdot 7 \stackrel{?}{=} -21$ $\qquad -21 = -21$ ✓
c. $\quad -4 = 4s$ $\quad 4s = -4 \quad \mid \div 4$ $\quad 4s/4 = -4/4$ $\quad \mathbf{s = -1}$ Check: $-4 \stackrel{?}{=} 4 \cdot (-1)$ $\qquad -4 = -4$ ✓	d. $\quad 72 = -6y \quad \mid \div (-6)$ $\quad 72/(-6) = -6y/(-6)$ $\quad -12 = y$ $\quad \mathbf{y = -12}$ Check: $72 \stackrel{?}{=} -6 \cdot (-12)$ $\qquad 72 = 72$ ✓

4.

a. $\quad -5q = -40 - 5$ $\quad -5q = -45 \quad \mid \div 5$ $\quad -5q/(-5) = -45/(-5)$ $\quad \mathbf{q = 9}$ Check: $-5 \cdot 9 \stackrel{?}{=} -40 - 5$ $\qquad -45 = -45$ ✓	b. $\quad 2 \cdot 36 = -6y$ $\quad 72 = -6y \quad \mid \div (-6)$ $\quad 72/(-6) = -6y/(-6)$ $\quad y = -12$ Check: $2 \cdot 36 \stackrel{?}{=} -6 \cdot (-12)$ $\qquad 72 = 72$ ✓
c. $\quad 3x = -4 + 3 + (-2)$ $\quad 3x = -3 \quad \mid \div 3$ $\quad 3x/3 = -3/3$ $\quad \mathbf{x = -1}$ Check: $3 \cdot (-1) \stackrel{?}{=} -4 + 3 + (-2)$ $\qquad -3 = -3$ ✓	d. $\quad 5 \cdot (-4) = -10z$ $\quad -20 = -10z \quad \mid \div (-10)$ $\quad -20/(-10) = -10z/(-10)$ $\quad 2 = z$ $\quad \mathbf{z = 2}$ Check: $5 \cdot (-4) \stackrel{?}{=} -10 \cdot 2$ $\qquad -20 = -20$ ✓

Multiplication and Division Equations, cont.

5.

a.	$\dfrac{x}{2} = -45 \qquad \mid \cdot 2$ $2 \cdot \dfrac{x}{2} = 2 \cdot (-45)$ $x = -90$ Check: $-90/2 \stackrel{?}{=} -45$ $\qquad -45 = -45$ ✓		b.	$\dfrac{s}{-7} = -11 \qquad \mid \cdot (-7)$ $-7 \cdot \dfrac{s}{-7} = -7 \cdot (-11)$ $s = 77$ Check: $77/(-7) \stackrel{?}{=} -11$ $\qquad -11 = -11$ ✓
c.	$\dfrac{c}{-7} = 4 \qquad \mid \cdot (-7)$ $-7 \cdot \dfrac{c}{-7} = -7 \cdot 4$ $c = -28$ Check: $-28/(-7) \stackrel{?}{=} 4$ $\qquad 4 = 4$ ✓		d.	$\dfrac{a}{-13} = -9 + (-11) \mid \cdot (-13)$ $-13 \cdot \dfrac{a}{-13} = -13 \cdot (-20)$ $a = 260$ Check: $260/(-13) \stackrel{?}{=} -9 + (-11)$ $\qquad -20 = -20$ ✓

6. a. Let x be the depth at which the shark is swimming. $x = -500/6$; $x = -83\ 1/3$.
   The shark is at a depth of $83\ 1/3$ feet below sea level.

   b. Let c be the total cost. $c/3 = \$21{,}200$; $c = \$63{,}600$. The total costs were \$63,200.

7.

a.	$\dfrac{1}{3}x = -15 \qquad \mid \cdot 3$ $3 \cdot \dfrac{1}{3}x = 3 \cdot (-15)$ $x = -45$ Check: $\dfrac{1}{3}(-45) \stackrel{?}{=} -15$ $\qquad -15 = -15$ ✓	b.	$-\dfrac{1}{6}x = -20 \qquad \mid \cdot (-6)$ $(-6) \cdot \left(-\dfrac{1}{6}\right)x = (-6) \cdot (-20)$ $x = 120$ Check: $-\dfrac{1}{6}(120) \stackrel{?}{=} -20$ $\qquad -20 = -20$ ✓	c.	$-\dfrac{1}{4}x = 18 \qquad \mid \cdot (-4)$ $-4 \cdot \left(-\dfrac{1}{4}\right)x = -4 \cdot 18$ $x = -72$ Check: $-\dfrac{1}{4}(-72) \stackrel{?}{=} 18$ $\qquad 18 = 18$ ✓
d.	$-2 = -\dfrac{1}{9}x \qquad \mid \cdot (-9)$ $-9 \cdot (-2) = -9 \cdot \left(-\dfrac{1}{9}\right)x$ $18 = x$ $x = 18$ Check: $-2 \stackrel{?}{=} -\dfrac{1}{9}(18)$ $\qquad -2 = -2$ ✓	e.	$-21 = \dfrac{1}{8}x \qquad \mid \cdot 8$ $8 \cdot (-21) = 8 \cdot \dfrac{1}{8}x$ $-168 = x$ $x = -168$ Check: $-21 \stackrel{?}{=} \dfrac{1}{8} \cdot (-168)$ $\qquad -21 = -21$ ✓	f.	$\dfrac{1}{12}x = -7 + 5$ $\dfrac{1}{12}x = -2 \qquad \mid \cdot 12$ $12 \cdot \dfrac{1}{12}x = 12 \cdot (-2)$ $x = -24$ Check: $\dfrac{1}{12}(-24) \stackrel{?}{=} -7 + 5$ $\qquad -2 = -2$ ✓

Word Problems, p. 57

1. Let s be one side of the square. Equation:
   $4s = 456$
   $s = 114$

   One side is 114 cm long.

2. Let s be the unknown side of the park. Equation:
   $62s = 4{,}588$
   $s = 74$.

   The other side is 74 feet long.

3. Let n be the number of boxes John bought. Equation:
   $15n = 165$
   $n = 11$.

   John bought 11 boxes of screws.

4. Let x be the original length of the candle. Equation:
   $x - 4 \cdot 2 = 6$
   $x = 14$

   Another possible equation is $2 \cdot 4 + 6 = x$.
   The candle was originally 14 cm long.

5. Let w be the weight of the baby dolphin. Equation:
   $(1/12\,w) = 15$
   $w = 180$

   The mother dolphin weighs 180 kilograms.

6.

Bar model:	Equation:
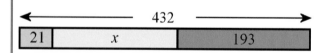 To find the value of x we need to subtract 21 and 193 from 432. $x = 432 - 21 - 193 = 218$	$21 + x + 193 = 432$ $x + 214 = 432 \quad \vert -214$ $x = 432 - 214$ $x = 218$

7.

Bar model:	Equation:
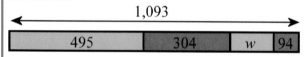 To find the value of w we need to subtract 495, 304, and 94 from 1,093. $w = 1{,}093 - 495 - 304 - 94 = 200$	$495 + 304 + w + 94 = 1{,}093$ $w + 893 = 1{,}093 \quad \vert -893$ $w = 1{,}093 - 893$ $w = 200$

Constant Speed, p. 60

1.

a. It will take the caterpillar 1 7/10 minutes, which is 1 min. 42 s to crawl 34 cm.		b. Father arrives 25 minutes later, which is at 8:05 a.m.	
$34 = 20t$	Flip the sides.	$20 = 48t$	Flip the sides.
$20t = 34$	Divide both sides by 20.	$48t = 20$	Divide both sides by 48
$\dfrac{20t}{20} = \dfrac{34}{20}$	The factors of 20 in numerator and denominator cancel.	$\dfrac{48t}{48} = \dfrac{20}{48}$	The factors of 48 in numerator and denominator cancel.
$t = 1\ 7/10$		$t = 5/12$	In minutes, 5/12 h = 25 min.

2. a. 35/60 = 0.583 hours.   b. 44/60 = 0.733 hours.   c. 2 + 16/60 = 2.267 hours.   d. 4 + 9/60 = 4.154 hours.

3. a. 2 h + 0.4 · 60 min = 2 h 24 min.   b. 0.472 · 60 min ≈ 28 min.
   c. 3 + 3/5 · 60 min = 3 h 36 min.   d. 16/50 · 60 min ≈ 19 min.

4. The bus can travel <u>272 km</u> in 4 hours and 15 minutes.
   Using the equation $d = vt$, we get d = 64 km/h · 4.25 h = 272 km.

5. Sam can run 2.4 miles in approximately <u>14 minutes</u>.
   Using the equation $d = vt$, we get 2.4 mi. = 10 mi/h · t, from which t = 2.4/10 h = 0.24 h = 14.4 minutes.

6. 4 hours 24 minutes.
   For the first half of the distance, we can write the equation, 180 mi. = 90 mi/h · t, from which t = 180/90 h = 2 h.
   For the second half of the distance, we get 180 mi. = 75 mi/h · t, from which t = 180/75 h = 2.4 h.
   In total, the train took 4.4 h = 4 h 24 min.

7. He can jog 2.5 miles.
   From the formula $d = vt$, we get the equation d = 6 mi/h · 25/60 h = 2.5 mi.

8. a. The duck flies 30 miles per hour.
   Since it flies 3 miles in 6 minutes, it will fly 30 miles in 60 minutes, which is 1 hour.
   Another way: $v = d/t$ = 3 mi/6 min = 3 mi/(6/60 h) = 3 mi/(1/10 h) = 30 mi/h.

   b. The lion runs 54 kilometers per hour.
   If it runs 0.9 km in 1 minute, then in 60 minutes, it runs 60 · 0.9 km = 54 km.
   Another way: $v = d/t$ = 0.9 km/1 min = 0.9 km/(1/60 h) = 54 km/h.

   c. Henry sleds down the hill 5/6 meters per second.
   $v = d/t$ = 75 m/1.5 min = 75 m/90 s = 15/18 m/s = 5/6 m/s

   d. Rachel swims 3/4 (0.75) kilometers per hour.
   $v = d/t$ = 400 m/32 min = 0.4 km/(32/60 h) = 0.4 km/(8/15 h) = 0.75 km/h.

9. a. His average speed going there was 67.2 kilometers per hour.
   It took him 2 h 14 min = 2 14/60 h ≈ 2.233 h to get there. His average speed was $v = d/t$ = 150 km/2.233 h ≈ 67.2 km/h.

   b. His average speed coming back was 78.2 kilometers per hour.
   It took him 1 h 55 min = 1 55/60 h ≈ 1.917 h to get there. His average speed was $v = d/t$ = 150 km/1.917 h ≈ 78.2 km/h.

   c. His overall average speed was 72.3 kilometers per hour.
   His total travel time was 4 h 9 min = 4 9/60 h = 4.15 h. His average speed was $v = d/t$ = 300 km/4.15 h ≈ 72.3 km/h.

10. It took Jake two hours to drive to his grandparents' place.
    It took Jake two hours and 18 minutes to drive home: t = 150 km/(65 km/h) ≈ 2.308 h ≈ 2 h 18 min.
    So, it took Jake <u>18 minutes longer</u> to drive home than to drive there.

11. We can calculate the average speeds of the two birds in miles per minute and then compare them.
    Seagull: v = 10 mi/24 min = 0.417 mi/min; eagle: 14 mi/30 min = 0.467 mi/min. The eagle is faster.

Constant Speed, cont.

12. a. For the first half of the trip, the train took 80 km/(60 km/h) = 8/6 h = 4/3 h = 1 h 20 min.
    For the second half of the trip, the train took 80 km/(240 km/h) = 1/3 h = 20 min.
    In total, it took 1 hour 40 minutes for the whole trip.

    b. Normally, it would have taken 160 km/(120 km/h) = 16/12 h = 4/3 h = 1 h 20 min for the whole trip.
    Notice that doubling the speed for the last half did not make up for the time that was lost traveling the first half at half the normal speed.

    c. Its average speed was $v = d/t$ = 160 km/(1 h 40 min) = 160 km/(5/3 h) = 96 km/h.

13. The first 1/3 of the trip: t = 1.5 km/(12 km/h) = 0.125 h = 7.5 min.
    The last 2/3 of the trip: t = 3 km/(15 km/h) = 3/15 h = 1/5 h = 12 min.
    It will take her a total of <u>19.5 minutes</u>.

14. No, you cannot. Essentially, walking half of the distance at half of your normal speed takes you the exact same time as walking the whole distance at your normal speed, so then it is not possible to make up for the lost time.

    Let's say that it is ten miles from your home to the pool.
    The first half takes you t = 5 mi/(3 mi/h) = 5/3 h = 1 h 40 min.
    The second half takes you t = 5 mi/(12 mi/h) = 5/12 h = 25 min.

    Walking with your normal speed, the trip would have taken t = 10 mi/(6 mi/h) = 5/3 mi/h = 1 h 40 min.

15. To make up for the lost time, the airplane should fly at 1,143 km per hour.

    In normal conditions, the airplane takes t = 1,600 km/(1,000 km/h) = 1.6 h = 96 min for the whole trip.
    In the first 40 minutes = 2/3 h of the trip, the airplane flies 2/3 h · 800 km/h = 533.33 km. This means there is a distance of 1600 km − 533 km = 1,067 km left of the trip.

    To make up for the lost time, the airplane has 96 min − 40 min = 56 min = 0.933 h to fly at a faster speed.
    Its average speed would need to be 1,067 km/0.933 h = 1,143 km per hour.

Two-Step Equations, p. 67

1.

a.
$5x + 2 = 67 \quad | -2$
$5x = 65 \quad | \div 5$
$x = 13$
Check:
$5 \cdot 13 + 2 \stackrel{?}{=} 67$
$65 + 2 \stackrel{?}{=} 67$
$67 = 67$ ✓

b.
$3y - 2 = 71 \quad | +2$
$3y = 73 \quad | \div 3$
$y = 73/3 = 24\ 1/3$
Check:
$3 \cdot (24\ 1/3) - 2 \stackrel{?}{=} 71$
$73 - 2 \stackrel{?}{=} 71$
$71 = 71$ ✓

c.
$-2x + 11 = 75 \quad | -11$
$-2x = 64 \quad | \div (-2)$
$x = -32$
Check:
$-2 \cdot (-32) + 11 \stackrel{?}{=} 75$
$64 + 11 = 75$ ✓

d.
$8z - 2 = -98 \quad | +2$
$8z = -96 \quad | \div 8$
$x = -12$
Check:
$8 \cdot (-12) - 2 \stackrel{?}{=} -98$
$-96 - 2 = -98$ ✓

2.

a.
$\dfrac{x+6}{5} = 14 \quad | \cdot 5$
$\dfrac{\cancel{5} \cdot (x+6)}{\cancel{5}} = 70$
$x + 6 = 70 \quad | -6$
$x = 64$
Check: $\dfrac{64 + 6}{5} \stackrel{?}{=} 14$
$\dfrac{70}{5} = 14$ ✓

b.
$\dfrac{x+2}{7} = -1 \quad | \cdot 7$
$\dfrac{\cancel{7} \cdot (x+2)}{\cancel{7}} = -7$
$x + 2 = -7 \quad | -2$
$x = -9$
Check: $\dfrac{-9 + 2}{7} \stackrel{?}{=} -1$
$\dfrac{-7}{7} = -1$ ✓

c.
$\dfrac{x-4}{12} = -3 \quad | \cdot 12$
$\dfrac{\cancel{12} \cdot (x-4)}{\cancel{12}} = -36$
$x - 4 = -36 \quad | +4$
$x = -32$
Check: $\dfrac{-32 - 4}{12} \stackrel{?}{=} -3$
$\dfrac{-36}{12} = -3$ ✓

d.
$\dfrac{x+1}{-5} = -21 \quad | \cdot (-5)$
$\dfrac{\cancel{-5} \cdot (x+1)}{\cancel{-5}} = 105$
$x + 1 = 105 \quad | -1$
$x = 104$
Check: $\dfrac{104 + 1}{-5} \stackrel{?}{=} -21$
$\dfrac{105}{-5} = -21$ ✓

Two-Step Equations, cont.

3.

a.
$$\frac{x}{10} + 3 = -2 \qquad |-3$$
$$\frac{x}{10} = -5 \qquad |\cdot 10$$
$$x = -50$$

Check: $\frac{-50}{10} + 3 \stackrel{?}{=} -2$
$-5 + 3 \stackrel{?}{=} -2$
$-2 = -2$ ✓

b.
$$\frac{x+3}{10} = -2 \qquad |\cdot 10$$
$$\frac{10 \cdot (x+3)}{10} = -20$$
$$x + 3 = -20 \qquad |-3$$
$$x = -23$$

Check: $\frac{-23+3}{10} \stackrel{?}{=} -2$
$\frac{-20}{10} \stackrel{?}{=} -2$
$-2 = -2$ ✓

4.

a.
$$\frac{x}{7} - 8 = -5 \qquad |+8$$
$$\frac{x}{7} = 3 \qquad |\cdot 7$$
$$x = 21$$

Check: $\frac{21}{7} - 8 \stackrel{?}{=} -5$
$3 - 8 \stackrel{?}{=} -5$
$-5 = -5$ ✓

b.
$$\frac{x-8}{7} = -5 \qquad |\cdot 7$$
$$x - 8 = -35 \qquad |+8$$
$$x = -27$$

Check: $\frac{-27-8}{7} \stackrel{?}{=} -5$
$\frac{-35}{7} \stackrel{?}{=} -5$
$-5 = -5$ ✓

5.

a.
$$1 - 5x = 2 \qquad |-1$$
$$-5x = 1 \qquad |\div (-5)$$
$$x = -1/5$$

Check: $1 - 5 \cdot (-1/5) \stackrel{?}{=} 2$
$1 + 1 \stackrel{?}{=} 2$ ✓

b.
$$12 - 3y = -6 \qquad |-12$$
$$-3y = -18 \qquad |\div (-3)$$
$$y = 6$$

Check: $12 - 3 \cdot 6 \stackrel{?}{=} -6$
$12 - 18 \stackrel{?}{=} -6$ ✓

Two-Step Equations, cont.

5.

c.	$10 = 8 - 4y$	$\mid -8$		d.	$7 = 5 - 3t$	$\mid -5$
	$-4y = 2$	$\mid \div (-4)$			$-3t = 2$	$\mid \div (-3)$
	$y = -1/2$				$t = -2/3$	
Check:	$10 \stackrel{?}{=} 8 - 4 \cdot (-1/2)$			Check:	$7 \stackrel{?}{=} 5 - 3 \cdot (-2/3)$	
	$10 \stackrel{?}{=} 8 + 2$ ✓				$7 \stackrel{?}{=} 5 + 2$ ✓	

6. $2x + 10 = 14$

7. $-9 - 3x = -8.3$

8.

a.
$$\frac{2x}{5} = 8 \quad \mid \cdot 5$$
$$\frac{5 \cdot 2x}{5} = 40$$
$$2x = 40 \quad \mid \div 2$$
$$x = 20$$

Check: $\frac{2 \cdot 20}{5} \stackrel{?}{=} 8$
$\frac{40}{5} \stackrel{?}{=} 8$
$8 = 8$ ✓

b.
$$\frac{3x}{8} = -9 \quad \mid \cdot 8$$
$$\frac{8 \cdot 3x}{8} = 8 \cdot (-9)$$
$$3x = -72 \quad \mid \div 3$$
$$x = -24$$

Check: $\frac{3 \cdot (-24)}{8} \stackrel{?}{=} -9$
$\frac{-72}{8} \stackrel{?}{=} -9$
$-9 \quad -9$ ✓

c.
$$15 = \frac{-3x}{10} \quad \mid \cdot 10$$
$$150 = -3x \quad \mid \div (-3)$$
$$-50 = x$$
$$x = -50$$

Check: $15 \stackrel{?}{=} \frac{-3(-50)}{10}$
$15 \stackrel{?}{=} \frac{150}{10}$
$15 = 15$ ✓

d.
$$2 - 4 = \frac{s + 4}{5} \quad \mid \cdot 5$$
$$10 - 20 = s + 4$$
$$-10 = s + 4 \quad \mid -4$$
$$-14 = s$$
$$s = -14$$

Check: $2 - 4 \stackrel{?}{=} \frac{-14 + 4}{5}$
$-2 = \frac{-10}{5}$ ✓

Two-Step Equations, cont.

8.

e.
$$\frac{x}{2} - (-16) = -5 \cdot 3$$
$$\frac{x}{2} + 16 = -15 \quad | -16$$
$$\frac{x}{2} = -31 \quad | \cdot 2$$
$$x = -62$$

Check: $\frac{-62}{2} - (-16) \stackrel{?}{=} -5 \cdot 3$
$-31 + 16 = -15$ ✓

f.
$$2 - 4p = -0.5 \quad | -2$$
$$-4p = -2.5 \quad | \div (-4)$$
$$p = 0.625$$

Check: $2 - 4(0.625) \stackrel{?}{=} -0.5$
$2 - 2.5 = -0.5$ ✓

Two-Step Equations: Practice, p. 72

1.

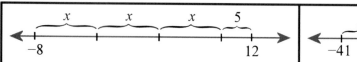

a.
$$-8 + 3x + 5 = 12$$
$$-3 + 3x = 12 \quad | +3$$
$$3x = 15 \quad | \div 3$$
$$x = 5$$

Check: $-8 + 3(5) + 5 \stackrel{?}{=} 12$
$-8 + 15 + 5 \stackrel{?}{=} 12$
$12 = 12$ ✓

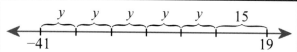

b.
$$-41 + 5y + 15 = 19$$
$$-26 + 5y = 19 \quad | +26$$
$$5y = 45 \quad | \div 5$$
$$y = 9$$

Check: $-41 + 5(9) + 15 \stackrel{?}{=} 19$
$-41 + 45 + 15 \stackrel{?}{=} 19$
$19 = 19$ ✓

2.

a.
$$2 - 5y = -11 \quad | -2$$
$$-5y = -13 \quad | \div (-5)$$
$$y = 13/5 = 2\ 3/5$$

Check: $2 - 5(2\ 3/5) \stackrel{?}{=} -11$
$2 - 13 \stackrel{?}{=} -11$
$-11 = -11$ ✓

b.
$$5y - 2 = -11 \quad | +2$$
$$5y = -9 \quad | \div 5$$
$$y = -9/5 = -1\ 4/5$$

Check: $5(-1\ 4/5) - 2 \stackrel{?}{=} -11$
$-9 - 2 \stackrel{?}{=} -11$
$-11 = -11$ ✓

Two-Step Equations Practice, cont.

3.

a.
$$\frac{2x}{7} = -5 \quad | \cdot 7$$
$$2x = -35 \quad | \div 2$$
$$x = -17\tfrac{1}{2}$$

Check: $\frac{2(-17\tfrac{1}{2})}{7} \stackrel{?}{=} -5$
$\frac{-35}{7} \stackrel{?}{=} -5$
$-5 = -5 \checkmark$

b.
$$\frac{x+2}{7} = -5 \quad | \cdot 7$$
$$x + 2 = -35 \quad | -2$$
$$x = -37$$

Check: $\frac{-37+2}{7} \stackrel{?}{=} -5$
$\frac{-35}{7} \stackrel{?}{=} -5$
$-5 = -5 \checkmark$

4.

a.
$$20 - 3y = 65 \quad | -20$$
$$-3y = 45 \quad | \div (-3)$$
$$y = -15$$

Check: $20 - 3(-15) \stackrel{?}{=} 65$
$20 + 45 \stackrel{?}{=} 65$
$65 = 65 \checkmark$

b.
$$6z + 5 = -2.2 \quad | -5$$
$$6z = -7.2 \quad | \div 6$$
$$z = -1.2$$

Check: $6(-1.2) + 5 \stackrel{?}{=} -2.2$
$-7.2 + 5 \stackrel{?}{=} -2.2$
$-2.2 = -2.2 \checkmark$

c.
$$\frac{t+6}{-2} = -19 \quad | \cdot (-2)$$
$$t + 6 = 38 \quad | -6$$
$$t = 32$$

Check: $\frac{32+6}{-2} \stackrel{?}{=} -19$
$\frac{38}{-2} \stackrel{?}{=} -19$
$-19 = -19 \checkmark$

d.
$$\frac{y}{6} - 3 = -0.7 \quad | +3$$
$$\frac{y}{6} = 2.1 \quad | \cdot 6$$
$$y = 12.6$$

Check: $\frac{12.6}{6} - 3 \stackrel{?}{=} -0.7$
$2.1 - 3 \stackrel{?}{=} -0.7$
$-0.7 = -0.7 \checkmark$

Two-Step Equations Practice, cont.

5.

a. Equation: Let s be the length of side for each of the three congruent sides. Then: $3s + 1.4\text{ m} = 7.1\text{ m} \quad \vert -1.4\text{ m}$ $3s = 5.7\text{ m} \quad \vert \div 3$ $s = 1.9\text{ m}$	Mental math/logical thinking: Subtract the given side length from the total length of the perimeter: 7.1 m − 1.4 m = 5.7 m. Then divide that result by 3 to get the length of any one of the three congruent sides: 5.7 m ÷ 3 = 1.9 m.
b. Equation: Let p be the price of one basket without any discounts. $6p - \$12 = \$46.80 \quad \vert + \$12$ $6p = \$58.80 \quad \vert \div 6$ $p = \$9.80$	Logical thinking: First add what you paid and the discount: $46.80 + $12 = $58.80. That is the cost of six baskets with the original price. Now, divide that by 6: $58.80 ÷ 6 = $9.80 to find the original cost of one basket.
c. Equation: Let x be the unknown number. $\dfrac{2}{5}x = 466 \quad \vert \cdot 5$ $2x = 2330 \quad \vert \div 2$ $x = 1{,}165$	Logical thinking: Since 466 is two fifths of the original number, divide 466 ÷ 2 = 233 to get one-fifth of the number. Then multiply that by five, and you have the original number: 5 · 233 = 1,165.

Growing Patterns 2, p. 76

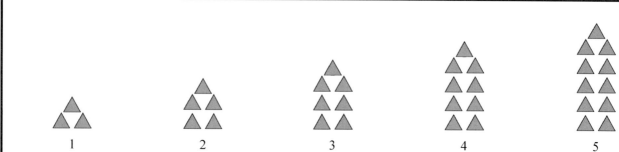

1. a. In each step, the pattern adds two triangles on the bottom.
   b. There will be 39 · 2 + 1 = 79 triangles in step 39.
   c. Answers may vary. Check the student's expression. It should be equivalent to $2n + 1$.
      For example, there will be $2n + 1$ or $(2n − 2) + 3$ triangles in step n.
   d. $2n + 1 = 311$
      $2n = 310$
      $n = 155$

2. a. In each step, the pattern adds one snowflake on the second and one on the third row.
   b. There will be (2 · 39) + 3 = 81 snowflakes in step 39.
   c. Answers may vary. Check the student's expression. It should be equivalent to $2n + 3$.
      For example, there will be $2n + 2 + 1$ or $2n + 3$ triangles in step n.
   d. $2n + 3 = 301$
      $2n = 298$
      $n = 149$

Growing Patterns 2, cont.

3.

1
2
3
...

Row	Flowers
1	3 + 1
2	3 + 2
3	3 + 3
4	3 + 4
5	3 + 5
6	3 + 6
7	3 + 7
8	3 + 8
n	3 + n

a. There will be $3 + n$ flowers in row n.
b. There will be $3 + 28 = 31$ flowers in the 28th row.
c. $3 + n = 97$
   $n = 94$
   There will be 97 flowers in row 94.

4. a. There are $2n + 1$ flowers in row n.
   b. $2n + 1 = 97$
      $2n = 96$
      $n = 48$

0
1
2
3

5. a. The total height of the building is $9n$ feet.
   b. The total height of the building is $9n + 2$ feet.
   c. (1) Using logical thinking: If the total height of the building is 164 ft, then the total height of all the stories is 162 ft. There are therefore $162 \div 9 = 18$ stories.

   (2) Using an equation: $9n + 2 = 164$
   $9n = 162$
   $n = 18$

6. a. $15n + 400$

   b. Equation: $15n + 400 = 970$
   $15n = 570$
   $n = 38$

   He should work 38 hours overtime.

Overtime hours	Total earnings	OR Total earnings
0	400 + 0	400
1	400 + 15	415
2	400 + 30	430
4	400 + 60	460
8	400 + 120	520
10	400 + 150	550
16	400 + 240	640
17	400 + 255	655
30	400 + 450	850
n	400 + 15n	400 + 15n

Puzzle corner:

   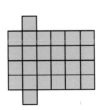 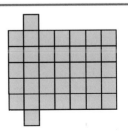

Step   1   2   3   4   5

One way to see this pattern is that it has one individual square at the top and one at the bottom, and in the middle it has a rectangle that is n squares high and $n + 2$ squares long. So in step n, the pattern has $n(n + 2) + 2$ squares.

In step 59, there will be $59(61) + 2 =$ **3,601 squares**.

Review 2, p. 80

1. a. $x = -13$ b. $x = 4$ c. $x = 10$ d. $z = 9$
   e. $x = -132$ f. $q = 120$ g. $c = -1{,}000$ h. $a = -105$

2. Equation: $3p = 837$. Solution: $p = 279$. One solar panel cost $279.

3. Equation: $s/7 = 187$. Solution: $s = 1{,}309$. Andrew's salary was $1,309.

4. Substituting the values given in the problem in the formula $d = vt$ gives us the equation $1.2 \text{ km} = 20 \text{ km/h} \cdot t$.
   Solution: $t = 1.2 \text{ km} / (20 \text{ km/h}) = 0.06 \text{ h} = 3.6 \text{ min} = 3 \text{ min. } 36 \text{ sec.}$

5. $v = d/t = 1.5 \text{ km}/3 \text{ min} = 1.5 \text{ km}/(3/60 \text{ h}) = 1.5 \text{ km}/(1/20 \text{ h}) = 30 \text{ km/h}$.

6. The first half: $t = 1 \text{ mi}/12 \text{ mph} = (1/12) \text{ h} = 5$ minutes. The second half: $t = 1 \text{ mi}/15 \text{ mph} = (1/15) \text{ h} = 4$ minutes. It will take Ed $5 + 4 = 9$ minutes to get to school.

Get Ready for Algebra: Grade 7 Alignment to the Common Core Standards

The table below lists each lesson and next to it the relevant Common Core Standard.

Lesson	page number	Standards
The Order of Operations	17	6.EE1. 6.EE2. 6.EE.6
Expressions and Equations	21	6.EE.2 6.EE.5 6.EE.6
Properties of the Four Operations	24	7.EE.1
Simplifying Expressions	28	6.EE.2 7.EE.1 7.EE.4
Growing Patterns 1	32	7.EE.2 7.EE.3 7.EE.4
The Distributive Property	35	7.EE.1 7.EE.2 7.EE.3 7.EE.4
Review 1	40	7.EE.1 7.EE.3
Solving Equations	42	6.EE.5 6.EE.7 7.EE.4
Addition and Subtraction Equations	49	7.EE.4
Multiplication and Division Equations	53	7.EE.4
Word Problems	57	7.EE.4
Constant Speed	60	7.RP.2.c 7.EE.3 7.EE.4
Two-step Equations	67	7.EE.4
Two-step Equations: Practice	72	7.EE.4
Growing Patterns 2	76	7.EE.2 7.EE.4
Review 2	80	7.EE.3 7.EE.4